MANGROSANA

MANGROSANA
FUTURE WORLD
RISING TIDES, SINKING ISLANDS & THE ROLE OF
MANGROVE TREES
***

PAPERBACK ISBN: 979-8-9892919-3-9
EPUB ISBN: 979-8-2237415-4-1
***

WRITTEN BY MARIA COWEN
PUBLISHED BY ROYAL HAWAIIAN PRESS
COVER ART BY TYRONE ROSHANTHA
RESEARCH & EDITORIAL ASSISTANCE: WARDA FIRDOUS
PUBLISHING ASSISTANCE BY DOROTA RESZKE
***

MANGROSANA

FUTURE WORLD

RISING TIDES, SINKING ISLANDS, & THE ROLE OF MANGROVE TREES

BY MARIA COWEN

RISING TIDES, SINKING ISLANDS, AND THE ROLE OF MANGROVE TREES

INTRODUCTION

According to van der Ploeg et al.'s in 2020 research climate change, including rising sea levels, poses a significant threat to global islands and their inhabitants. The essay explores the impact of sea level rise, the fragility of island habitats, and the crucial role of mangrove trees in protecting these ecosystems. It underscores the complex relationship between rising seas and disappearing islands, emphasizing the vital role of mangrove trees in maintaining these unique ecosystems.

ISLANDS AT RISK: RISING SEAS

Numerous Pacific or Caribbean islands' peace is in jeopardy due to the huge increase in sea levels brought on by climate change. This is mostly caused by melting glaciers and polar ice caps, as well as the thermal expansion of saltwater brought on by temperature increases. The results are disastrous, especially for low-lying island countries, underscoring the urgent need for action to lessen the catastrophic impacts of climate change. Rising tides may have an especially negative effect on low-lying islands, which are sometimes just a few metres above sea level. These islands are home to diverse human groups with rich cultural histories in addition to distinctive flora and wildlife (van der Ploeg et al., 2020).

SEA-LEVEL RISE IMPACTS ON COASTLINES

Several disastrous effects emerge when sea levels rise (Ellison, 2014; Thakur et al., 2021):

1. Coastal Erosion: Due to increasing tides and storm surges, coastal erosion is becoming a greater hazard to shorelines, jeopardising the fragile ecosystems that rely on them. This relentless process consumes land, threatens homes and infrastructure, and interferes with the socioeconomic system. The biological effects of erosion are severe, upsetting ecosystems and testing the adaptability of coastal plants and wildlife.

2. Saltwater Intrusion: Seawater seeping into freshwater aquifers poses a significant threat to human survival and agricultural output. Contaminated drinking water for coastal towns threatens public health and water quality. Soil composition is disrupted when seawater seeps onto agricultural land, leading to infertile soil and preventing crop development. This simultaneous attack on water resources and arable land threatens food security, necessitating adaptable solutions to protect freshwater supplies and agricultural viability in coastal areas.

3. Habitat Loss: Wetlands and mangrove forests on islands are crucial ecosystems for marine life, supporting various species. Wetlands provide breeding and feeding grounds for various species due to their unique land-water combination. Mangrove forests provide natural habitats for aquatic species, providing protection and food. However, the loss of these ecosystems threatens the delicate biological balance, fisheries, and way of life for numerous populations relying on these resources.

4. Displacement of Communities: Island communities face significant challenges due to land submergence under escalating waves, leading to the displacement of their ancestral homes. This upheaval disrupts communal norms, weakens ties between generations, and signifies the disappearance of deeply ingrained traditions and practices. Economically, relocation disrupts local businesses and means of support, leaving communities to face the difficulties of reconstructing in other locations.

THE MANGROVE SOLUTION

Mangrove trees emerge as a beacon of hope and perseverance in the midst of these severe conditions. Mangroves are uniquely equipped to survive at the dynamic land-sea interface. They have exceptional adaptations that allow them to resist saltwater, high winds, and tidal variations (McKee et al., 2007). Mangroves are crucial in the battle against rising sea levels for the following reasons (Bera et al., 2021):

1. Coastal Protection: With its complex web of roots, mangroves protect the beach from erosion by anchoring it in a strong embrace. Their sinuous tendrils intertwine like the armoury of nature, steadily reducing the force of waves and surges to create an impenetrable barrier that stands sentinel against the advancing tides. In their unwavering defence, they provide protection, shielding the priceless land beyond the raging powers of the sea.

2. Sediment Trapping: The intricate root systems of mangroves trap sediments carried by water, helping to build and maintain the elevation of islands. This is especially crucial in regions where rising sea levels threaten to submerge entire islands.

3. Habitat Preservation: Mangrove ecosystems provide crucial homes for a wide range of species, including important creatures such as fish, crabs, and bird fauna. These green areas function as crucial breeding grounds for commercially important fish species, supporting fisheries that play a vital role in ensuring food security for those living along the coast. Amidst the complex web of ecological interactions, mangrove forests play a crucial role as essential guardians, ensuring the preservation of biodiversity and the provision of resources for human societies beyond their ecological importance.

4. **Carbon Sequestration:** On a global scale, mangroves are among carbon sequestration champions. They have a remarkable potential to accumulate significant amounts of carbon in both their biomass and the underground soils they live in. In the larger effort to lessen the effects of climate change, the preservation of mangrove ecosystems stands out as a powerful solution. It is a crucial route for reducing the rising trend of global carbon emissions.

5. **Community Livelihoods:** Mangroves are a vital source of livelihood for many island communities, contributing in a variety of ways to the residents' economic well-being. They provide essential materials for preserving traditional methods of construction, using medicine, and providing food. These communities serve as advanced examples of environmentally friendly cooperation with the environment in their symbiotic interaction with mangroves. This delicate balance is crucial to these communities' long-term survival.

CONCLUSION

The preservation of islands is no longer only about their aesthetic value in an era of worsening climate change; it has become an imperative of cultural legacy, ecological variety, and unyielding resilience. Mangrove trees, situated prominently, serve as resilient friends in the effort to combat the encroachment of elevated sea levels, providing a promising prospect for the conservation of these priceless sanctuaries. Acknowledging the limitations of mangroves in addressing the diverse array of issues encountered by islands, it is indisputable that they represent an essential element within a holistic approach to adapt to and mitigate the consequences of climate change effectively. Through the deliberate cultivation of mangrove forests, individuals engage in a transforming endeavour that serves to strengthen not only the geographical perimeters of these islands but also fosters a deep-rooted feeling of hope, revitalisation, and the lasting heritage of these idyllic locations for future generations.

REFERENCES

Bera, A., Taloor, A. K., Meraj, G., Kanga, S., Singh, S. K., Đurin, B., & Anand, S. (2021). Climate vulnerability and economic determinants: Linkages and risk reduction in Sagar Island, India; A geospatial approach. Quaternary Science Advances, 4, 100038. https://doi.org/10.1016/j.qsa.2021.100038

Ellison, J. C. (2014). Vulnerability assessment of mangroves to climate change and sea-level rise impacts. Wetlands Ecology and Management, 23(2), 115–137. https://doi.org/10.1007/s11273-014-9397-8 McKee, K. L., Cahoon, D. R., & Feller, I. C. (2007). Caribbean mangroves adjust to rising sea level through biotic controls on change in soil elevation. Global Ecology and Biogeography, 16(5), 545–556. https://doi.org/10.1111/j.1466-8238.2007.00317.x

Thakur, S., Mondal, I., Bar, S., Nandi, S., Ghosh, P. B., Das, P., & De, T. K. (2021). Shoreline changes and its impact on the mangrove ecosystems of some islands of Indian Sundarbans, North-East coast of India. Journal of Cleaner Production, 284, 124764. https://doi.org/10.1016/j.jclepro.2020.124764 van der Ploeg, J., Sukulu, M., Govan, H., Minter, T., & Eriksson, H. (2020). Sinking Islands, Drowned Logic; Climate Change and Community-Based Adaptation Discourses in Solomon Islands. Sustainability, 12(17), 7225. https://doi.org/10.3390/su12177225

PART I

Chapter 1:
OVERVIEW AND OBJECTIVES

Mangroves are a unique and essential habitat that is found in the center of the world's different coastal ecosystems. The ecological and socioeconomic balance of tropical and subtropical locations across the globe is greatly influenced by these extraordinary coastal forests, which are characterized by salt-tolerant trees and shrubs. Mangroves are crucial for the sustainability of many coastal settlements as well as the survival of various species. This introductory chapter lays the groundwork for the exploration of mangroves and their crucial function in the sustainability of islands.

1.1: THE MANGROVE ECOSYSTEM

The very intricate and dynamic ecosystems known as **mangroves,** sometimes known as "the nurseries of the sea," flourish in the area where land and water meet. The red mangrove (Rhizophora mangle), black mangrove (Avicennia germinans), and white mangrove (Laguncularia racemosa) are only a few of the distinctive species that define them. These species have developed impressive adaptations to survive in the harsh intertidal zone, which includes varying salinity, a lack of oxygen, and high temperatures.

An intricate web of life abounds in this setting. Mangroves provide homes, nesting areas, and food sources for many fish, crab, mollusk, and bird species. Juvenile fish find refuge in the complicated root systems of mangrove plants, shielded from more giant predators. These woods are essential for migrating birds as well because they provide necessary rest breaks for their lengthy oceanic migrations.

1.2: THE GLOBAL SIGNIFICANCE OF MANGROVES

Mangroves are crucial not just environmentally but also economically and culturally. Mangrove habitats are crucial to the survival of coastal people across the globe. Farmers find subsistence in the lush mudflats, while fishermen depend on the productive fishing grounds produced by mangroves. Additionally, these beautiful coastal settings often support a thriving tourist industry, giving the locals job possibilities.

Mangroves play a crucial role in mitigating the impacts of climate change. These entities are considered to be highly efficient carbon sinks on a global scale since they can extract significant amounts of carbon dioxide from the Earth's atmosphere. Protecting mangrove forests is crucial for slowing global warming and lowering carbon dioxide emissions[1].

1.3 AIMS OF THIS BOOK

The main aim of the book "Introduction to Mangroves and Island Sustainability" are as follows:

---

[1] Daniel M. Alongi, "Global Significance of Mangrove Blue Carbon in Climate Change Mitigation," *Sci* 2, no. 3 (August 21, 2020): 67, https://doi.org/10.3390/sci2030067.

1: **Educational Awareness:** One of the primary goals of this work is to raise students' awareness of the importance of mangroves as an ecosystem by explaining the unique adaptations of the species that live there and showcasing the wide variety of organisms that call this ecosystem home.

2. **Long-term viability:** To examine how mangroves balance economic growth and environmental protection to sustain an island.

3: **Sustainability:** To understand the Mangrove ecological restoration is crucial for future generations.

4. **Local Community Engagement:** To understand Mangrove protection, sustainable usage, and social and economic issues.

5: **Global Perspective:** This book uses case studies from throughout the globe to illustrate the significance of mangroves and their difficulties.

6. **Climate Change Mitigation:** Mangrove carbon sinks and adaptation.

7. **Interdisciplinary Approach:** To promote interdisciplinary thinking and cooperation among researchers, decision-makers, environmentalists, and local populations in tackling issues relating to mangroves.

The following chapters will take us on an exploration of discovery as we dig into the subtle beauty and relevance of mangroves in the context of island sustainability. Together, we will learn more about the urgent problems these ecosystems are now experiencing and investigate creative solutions targeted at protecting and restoring these coastal beauties for future generations.

Chapter 2:

UNDERSTANDING THE IMPORTANCE OF MANGROVES FOR ISLAND ECOSYSTEMS

2.0: INTRODUCTION TO MANGROVES AND ISLAND SUSTAINABILITY

It is essential to launch an in-depth inquiry into the fundamental causes driving the enormous significance of these coastal ecosystems in order to begin a thorough assessment of the critical role that mangroves play in fostering sustainability on islands. The biological intricacies that make mangroves essential elements in the complex global network of island ecosystems will be examined in-depth in the next chapter. A variety of marine species find critical homes in the complicated root systems of coastal plants, which also play a significant role in maintaining shorelines. We may better comprehend how these luxuriant ecosystems contribute to the general health and balance of our planet by looking at the intricate network of biological interaction. We can learn a lot about the substantial impacts that mangroves have on the larger ecological landscape by performing a detailed examination of the symbiotic relationships between mangroves and nearby ecosystems. Additionally, we will look at the less well-known but no less critical roles they play, including the filtration and purification of coastal waters and their contribution to nitrogen cycling. We will have a deeper understanding of the priceless services that mangroves provide as the process of discovering these natural treasures progresses, establishing the groundwork for a deeper understanding of their critical role in the sustainability of islands.

2.1: COASTAL GUARDIANS

Mangroves are essential coastal ecosystems with unique tree species that can withstand salt. They serve as protective barriers against natural processes, such as tides and waves, and absorb energy from storms. Their complex root structure impedes soil erosion and protects shoreline stability. Salt marshes provide a critical barrier against erosion and storm surges, ensuring human safety and protecting coastal infrastructure. Mangroves also play a double role in maintaining the balance between coastal ecosystems and the societies reliant on them. Their dual responsibility highlights the importance of these ecosystems in maintaining the delicate balance between coastal ecosystems and their inhabitants[2].

2.2: BIODIVERSITY HOTSPOTS

Mangrove habitats are known for their rich biodiversity, making them significant biodiversity hotspots. These ecosystems serve as refuges and breeding grounds for various marine organisms, with the complex interconnection of subterranean root systems and submerged ecological environments providing sanctuary for a variety of creatures. Young fish and crabs thrive in the protective sanctuary provided by mangrove roots, while bird species like herons, egrets, and kingfishers find respite above the water's surface. These ecosystems provide nourishment and habitats for these organisms, ensuring the perpetuation of their populations. The bird population relies heavily on the abundance of insects and invertebrates in the mangrove ecosystem. The complex choreography of existence illustrates the interconnectedness of these

---

[2] Raymond D. Ward et al., "Impacts of Climate Change on Mangrove Ecosystems: A Region by Region Overview," *Ecosystem Health and Sustainability* 2, no. 4 (April 2016): e01211, https://doi.org/10.1002/ehs2.1211.

ecological systems, with every species playing a vital role in maintaining biodiversity equilibrium[3].

2.3: NUTRIENT CYCLING

Mangroves play a crucial role in the nutrient cycle of coastal ecosystems by converting organic materials into nutritious food. The gradual descent of leaves and debris from mangrove trees into the brackish water initiates a process of decomposition, resulting in a broth rich in nutrients. This organic mixture serves as a significant source of nutrition for filter-feeding species, forming the foundation for a complex ecological network. Mangroves also serve as resilient protectors of water quality, intercepting sediments and acting as innate filtering mechanisms. They effectively remove high concentrations of heavy metals and surplus nutrients, preserving the adjacent coastal ecosystem and protecting coral reefs and seagrass beds downstream. Their nutrient cycling abilities sustain the vitality and complexity of coastal ecosystems, ensuring the cleanliness and soundness of their local habitat and the broader marine ecosystem they inhabit[4].

---

[3] Stefanie M. Rog, Rohan H. Clarke, and Carly N. Cook, "More than Marine: Revealing the Critical Importance of Mangrove Ecosystems for Terrestrial Vertebrates," ed. Robert Cowie, *Diversity and Distributions* 23, no. 2 (November 23, 2016): 221–30, https://doi.org/10.1111/ddi.12514.

[4] Alongi D. et al., "Below-Ground Nitrogen Cycling in Relation to

2.4: CARBON SEQUESTRATION

Mangroves are crucial in combating climate change by acting as natural carbon sequestration reservoirs. Their submerged root systems gather and retain organic waste, locking carbon in a hidden area, preventing its release and attenuating its impact on the Earth's atmosphere. Conservation and rehabilitation of mangrove ecosystems are a promising approach to combating climate change. These coastal habitats, also known as "blue carbon" ecosystems, have the ability to mitigate greenhouse gas levels. By preserving these ecosystems, we contribute to reducing greenhouse gas concentrations and promoting global climate mitigation efforts. Mangroves provide a natural remedy to climate change, highlighting their role as guardians of the planet's ecological balance[5].

2.5: SUSTAINING ISLAND LIVELIHOODS

Mangroves are vital for island populations due to their abundant coastal resources, which support

Net Canopy Production in Mangrove Forests of Southern Thailand," *Marine Biology* 140, no. 4 (April 1, 2002): 855–64, https://doi.org/10.1007/s00227-001-0757-6.

[5] Daniel M Alongi, "Carbon Sequestration in Mangrove Forests," *Carbon Management* 3, no. 3 (June 2012): 313–22, https://doi.org/10.4155/cmt.12.20.

sectors like fishing, aquaculture, and eco-tourism. The sustainable extraction of marine organisms like fish, crabs, and mollusks generates cash for these communities, supporting their livelihoods and enhancing local economic conditions. Mangrove habitats also attract tourists, generating revenue for local communities. To preserve these resources, island communities should prioritize environmentally friendly approaches and protect mangrove environments.[6].

2.6: A HOLISTIC PERSPECTIVE

In order to fully comprehend the significant role of mangroves in island ecosystems, it is essential to adopt a comprehensive and integrated viewpoint. These natural phenomena function as protective barriers along coastlines, areas of high biodiversity, agents of nutrient cycling, and reservoirs for carbon storage. As we go through the following chapters, we will undertake an exploration aimed at understanding the intricate balance between human activity, conservation efforts, and the sustainable governance of mangrove ecosystems. The significance of this research lies in its ability to

---

[6] Nur Fatin Nabilah Ruslan et al., "Mangrove Ecosystem Services: Contribution to the Well-Being of the Coastal Communities in Klang Islands," *Marine Policy* 144 (October 2022): 105222, https://doi.org/10.1016/j.marpol.2022.105222.

provide a comprehensive comprehension, hence enabling the formulation of a strategic trajectory towards a mutually beneficial cohabitation with these essential coastal ecosystems. Recognizing the importance of mangroves in maintaining island life, we embark on a joint mission to ensure the sustainability of these ecosystems and the well-being of the people who depend on them for generations to come.

Chapter 3:

CHALLENGES FACED BY SUBMERGING ISLANDS IN THE PACIFIC AND CARIBBEAN

Islands located in the Pacific and Caribbean areas are confronted with a multitude of issues, including the progressive submergence caused by the escalation of sea levels. This chapter aims to examine the intricate problems faced by these susceptible islands, with a specific emphasis on the convergence of environmental, social, and economic concerns. In addition to this, we will talk about the role mangroves play in reducing the impact of these problems.

3.1: RISING SEA LEVELS: AN IMMINENT THREAT

The Pacific and Caribbean islands are facing significant challenges due to climate change, particularly in coastal regions. The Intergovernmental Panel on Climate Change (IPCC) warns that these islands are experiencing a faster increase in sea levels than the world average. This is primarily due to thermal expansion in salt water and polar ice cap melting. This is causing coastal erosion, saltwater infiltration into freshwater reserves, and increased storm surges.

The increasing sea levels pose risks to island communities, who have lived in these environments for generations. The gradual advance of oceanic tides not only erodes sandy coastlines but also undermines human sustenance, traditional practices, and cultural legacy. As time progresses, the struggle to adapt and strengthen in response to this climatic disruption becomes more challenging as these resilient island nations find innovative ways to meet the imminent difficulties.[7].

3.2: VULNERABLE ECOSYSTEMS

The National Oceanic and Atmospheric Administration (NOAA) warns that increasing sea levels and ocean acidification threaten the Pacific and Caribbean islands, which are home to complex

---

[7] Richard H. Moss et al., "The next Generation of Scenarios for Climate Change Research and Assessment," *Nature* 463, no. 7282 (February 2010): 747–56, https://doi.org/10.1038/nature08823.

marine ecosystems such as coral reefs, seagrass beds, and mangrove forests. The rich biological diversity and ever-changing conditions of coral reefs make them especially susceptible to environmental damage. The phenomena of coral bleaching, which results in the paleness and debilitation of corals, is instigated by the escalation of water temperatures. This spoils their beauty and upsets the delicate balance of life that relies on these bright structures. The collapse of coral reefs has wide-ranging consequences that extend beyond the visible aspects of these species' struggle to adapt. The fisheries industry, which is strongly dependent on the abundant yields provided by coral reefs, encounters substantial obstacles. The complex interplay of marine organisms, which relies on the presence of these coral reefs, is disrupted, with significant repercussions across the more comprehensive ecological system. Seagrass beds and mangrove forests are facing threats due to ocean acidification and rising sea levels. These ecosystems provide shelter and nourishment for marine organisms but are also vulnerable. The ecological disruption affects island inhabitants and their species. The delicate balance between nature and human civilization calls for cooperation to safeguard these aquatic beauties for future generations[8].

[8] Elizabeth DeLoughrey and Tatiana Flores, "Submerged Bodies," *Environmental Humanities* 12, no. 1 (May 1, 2020): 132–66, https://doi.org/10.1215/22011919-8142242.

3.3: LOSS OF LAND AND HABITATS

Sea levels are transforming the shorelines of islands, leading to coastal erosion and flooding, causing land and habitat loss, particularly the depletion of mangrove forests. These forests are crucial for maintaining the stability of terrestrial and marine ecosystems, providing refuge and nourishment for various species. However, the retreat of mangrove forests due to advancing waves threatens the resilience of the populations living on these islands, as their settlements are more vulnerable to storms and tempests. The delicate balance between human communities and coastal environments is at risk, emphasizing the need for collective measures. The United Nations Environment Programme (UNEP) highlights the importance of mangroves and their immediate challenges from sea-level rise and habitat loss. As society faces increasing sea levels, it becomes our responsibility to protect these ecosystems and human settlements that depend on their protection. By implementing coordinated efforts and dedication to preservation, we can establish a future where terrestrial areas and means of subsistence flourish in symbiosis with the environment[9].

---

[9] C. R. Maag, "National Oceanic and Atmospheric Administration (NOAA) Contamination Monitoring Instrumentation," *NASA ADS* 216 (January 1, 1980): 87, https://ui.adsabs.harvard.edu/abs/1980SPIE..216...87M/abstract.

3.4: FRESHWATER SCARCITY

Pacific and Caribbean islands are facing freshwater shortages due to rising sea levels, with saltwater infiltrating valuable aquifers and disrupting freshwater supplies. Groundwater is crucial for daily necessities and agricultural pursuits, but seawater intrusion leads to unsuitable groundwater. Communities struggle to find alternative water sources, leading to desalination as an expensive and resource-intensive process. The World Bank's research emphasizes the urgent issue of freshwater shortages and saltwater intrusion, emphasizing the need for global initiatives to help these islands negotiate freshwater shortages and implement environmentally conscious strategies to protect human entitlement to uncontaminated and sustainable freshwater resources[10].

3.5: COASTAL COMMUNITIES AT RISK

Climate change is causing coastal communities in the Pacific and Caribbean to face a double challenge: increased susceptibility to severe weather events and land loss, and a radical shift in lifestyle. Displacement is a significant issue, as families are

---

[10] Bouillon, S., Rivera-Monroy, V., Twilley, R. R., & Kairo, J. G. (2009). Mangroves. *The management of natural carbon sinks in coastal ecosystems.*, 13-22.

forced to leave their ancestral homes, affecting their cultural heritage and traditional practices. Climate-induced migration also leads to erosion of cultural identity and disconnection from ancestral territories and natural resources. As time progresses, the challenge to maintain existence becomes more pronounced, with the fragile balance of their mode of living becoming increasingly uncertain. The United Nations Human Rights Council platform is promoting the voices of marginalized populations advocating for their rights and dignity in response to climate-related adversities. Collaborative efforts and support are crucial for these coastal towns to endure and persist[11].

3.6: MANGROVES AS NATURAL DEFENDERS

Mangrove trees, found along fragile islands' shorelines, are crucial in fighting coastal erosion due to their complex root systems that create an impenetrable barrier against waves and tides. These trees also retain sediments, trapping them as water passes through them, helping restore degraded land and shorelines. This sedimentation aids in the gradual building of land, increasing the stability of at-risk islands. Credible sources like the Food and Agriculture Organization (FAO) and Wetlands

---

[11] Mark Spalding, Mami Kainuma, and Lorna Collins, *World Atlas of Mangroves* (Routledge, 2010).

International emphasize the importance of mangroves in strengthening coastal communities' resilience. Vulnerable islands can take the first steps towards sustainable adaptation and protecting their valuable coastal assets by utilizing the mangrove forests on their shores[12].

3.7 CONCLUSION

The problems encountered by submerged islands in the Pacific and Caribbean regions are formidable and intricately linked. The pressing need for immediate action is evident due to the rising sea levels, degradation of ecosystems, lack of freshwater resources, and the heightened vulnerability faced by coastal people. Mangroves provide a promising prospect by offering nature-based solutions to address and alleviate a portion of these difficulties. In the following chapters, we will explore practical approaches to effectively use mangroves in the endeavour to achieve sustainable development on islands.

In order to effectively tackle these intricate problems, it is essential to prioritize multidisciplinary teamwork, well-informed

---

[12] Bernhard Lehner and Petra Döll, "Development and Validation of a Global Database of Lakes, Reservoirs and Wetlands," *Journal of Hydrology* 296, no. 1 (August 20, 2004): 1–22, https://doi.org/10.1016/j.jhydrol.2004.03.028.

policymaking, and active involvement with the community. The future of these distinct but fragile island countries may be made more sustainable and resilient if we all work toward that goal together.

PART II:
MANGROVE ECOLOGY AND TYPES

$\mathcal{C}$hapter 4:
EXPLORING THE DIFFERENT TYPES OF MANGROVE SPECIES

4.0: MANGROVE ECOLOGY AND TYPES

Mangroves are essential coastal ecosystems that flourish in tropical and subtropical regions worldwide. These ecosystems exhibit a broad array of plant and animal species, resulting in a sanctuary of high biodiversity. The value of these ecosystems goes beyond their rich biodiversity, including the provision of a diverse range of ecosystem services that are crucial for both natural ecosystems and human societies. In the context of this chapter, we undertake an enlightening exploration of the diverse array of mangrove species. This study contains the geographic extent, ecological value, and crucial statistical data of coastal forests, providing insights into their significant importance.

4.1: INTRODUCTION TO MANGROVE SPECIES

A wide variety of tree and plant species thrive inside the complex mangrove forests, well adapted to the rigorous crucible of salty and brackish waters. These species have evolved special adaptations throughout evolution that enable them to overcome obstacles provided by high salinity, the rhythmic ebb and flow of tides, and oxygen-depleted soils. Surprisingly, there are more than 100 different species of mangroves present on the whole planet. Despite this astounding variety, they may be divided into four main kinds because of the unique interactions between their physical characteristics and their precisely adjusted ecological responsibilities[13].

Key Statistics:

Number of mangrove species worldwide: Over 100

Main types of mangroves: Rhizophora, Avicennia, Sonneratia, and Laguncularia[14]

---

[13] Shekhar R. Biswas et al., "A Unified Framework for the Restoration of Southeast Asian Mangroves—Bridging Ecology, Society and Economics," *Wetlands Ecology and Management* 17, no. 4 (August 9, 2008): 365–83, https://doi.org/10.1007/s11273-008-9113-7.

[14] Mangrove Action Project

4.2: RHIZOPHORA MANGROVES

Rhizophora serves as a botanical representation of the dynamic and important mangrove ecosystems found all over the globe. Its broad distribution in coastal areas attests to its exceptional capacity to adapt to difficult environmental circumstances. The complicated network of prop roots that grow from the stem and arch above the waterline is what distinguishes Rhizophora from other plants. This one-of-a-kind adaption is a wonder of natural engineering since it increases the tree's resistance to the strong pressures of coastal tides while simultaneously providing structural support. These roots act as a fortress, keeping the mangrove in its current location and allowing nutrients and oxygen to be exchanged between the tree and its brackish aquatic surroundings. Rhizophora is at the vanguard of nature's defense against coastal erosion and storm surges as a result of its close ties to the intertidal zone, highlighting its crucial role in maintaining the delicate balance of coastal ecosystems.

Rhizophora's preference for living in the intertidal zone attests to its outstanding resilience to the difficulties posed by frequent tidal flooding. To adapt to the dynamic variations in water levels that characterize coastal habitats, these mangroves have developed complex processes. For example, their salt-secreting glands allow them to flourish in brackish environments by filtering out more salt, allowing them to survive when other terrestrial plants would wither. Additionally, a variety of marine and bird species may find shelter and hatching grounds under the thick canopy of Rhizophora, which promotes a rich biodiversity inside its beautiful embrace. Rhizophora mangroves are important for their biological value. Still, they are also essential for reducing the effects of climate change because they absorb carbon dioxide from the air and serve as a natural barrier against storm surges. Because of its crucial function in preserving our coastal ecosystems, Rhizophora is a living example of the wonders of evolution[15].

Key Statistics:

Number of Rhizophora species: Approximately 40

Global distribution: Found in tropical and subtropical regions worldwide

Ecological significance: Stabilize coastlines, provide habitat for numerous species, and offer valuable wood products[16].

[15] A E Lugo and S C Snedaker, "The Ecology of Mangroves," *Annual Review of Ecology and Systematics* 5, no. 1 (November 1974): 39–64, https://doi.org/10.1146/annurev.es.05.110174.000351.
[16] Smithsonian Marine Station

4.3: AVICENNIA MANGROVES

The Avicennia genus of mangroves is a crucial part of coastal ecosystems due to its unique characteristics. Pneumatophores, or aerial roots, help the tree absorb air, which is essential in waterlogged soils. These roots facilitate gas exchange between the plant and its aquatic surroundings, allowing Avicennia to thrive in conditions that would kill most terrestrial plants. Avicennia can survive in the intertidal zone despite constant water levels, stabilizing coastal ecosystems and providing habitat for marine and avian species. Its complex root structure aids in breathing, offers a firm footing resistant to erosion and tides, and contributes to the ecosystem's health by filtering and purifying coastal waters and cycling nutrients[17].

Key Statistics:

Number of Avicennia species: Approximately 8

Global distribution: Found in tropical and subtropical regions, with a focus on the Indo-Pacific

Ecological significance: Improve water quality, provide coastal protection, and support fisheries

---

[17] K. Kathiresan, "Mangroves: Types and Importance," *Mangroves: Ecology, Biodiversity and Management*, 2021, 1–31, https://doi.org/10.1007/978-981-16-2494-0_1.

4.4: SONNERATIA MANGROVES

Sonneratia mangroves, coastal vegetation, survive by producing viviparous propagules. In intricate intertidal environments, Sonneratia survives by establishing new plants from these propagules. They preserve shorelines and sustain species because of their water resistance and low tide exposure. Their salt excretion mechanisms let them survive excessive salinity. Sonneratia mangroves protect against erosion, provide wood and traditional cures, and spawn marine creatures[18].

Key Statistics:

Number of Sonneratia species: Approximately 7

Global distribution: Predominantly found in the Indo-West Pacific region

Ecological significance: Important habitat for birds and crabs, and involved in sediment stabilization[19].

---

[18] F. Blasco, M. Aizpuru, and C. Gers, "Depletion of the Mangroves of Continental Asia," *Wetlands Ecology and Management*, 2004, https://doi.org/10.1023/A:1011169025815.

[19] Michael Sievers et al., "Integrating Outcomes of IUCN Red List of Ecosystems Assessments for Connected Coastal Wetlands," *Ecological Indicators* 116 (September 1, 2020): 106489, https://doi.org/10.1016/j.ecolind.2020.106489.

4.5: LAGUNCULARIA MANGROVES

Along the shore, mangroves of the genus Laguncularia are renowned for their adaptability and variety. They thrive in estuaries and other places where freshwater and saltwater mix, where the water's salinity is quite low. These creatures are crucial to preserving ecological harmony and provide habitat to marine and avian species. Their vivid green leaves and white bark serve as ecological indicators that show their ability to survive in challenging conditions. Laguncularia mangroves are essential to the health of coastal ecosystems because they act as a natural barrier against erosion and storm surges. They are a tremendous benefit to human settlement because their deep root systems minimise soil erosion and loss near the beach. Laguncularia species are essential for maintaining the balance of the ecosystem by filtering and purifying coastal waters as well as cycling nitrogen. Their great range of adaptability and variability draw attention to their important function in coastal ecosystems across the globe.

Key Statistics:

Number of Laguncularia species: Approximately 1 (Laguncularia racemosa)

Global distribution: Found in the Americas, including the Gulf of Mexico and the Caribbean

Ecological significance: Contribute to sediment trapping, shoreline protection, and bird habitat[20].

4.6: OTHER MANGROVE SPECIES

Although the aforementioned mangrove kinds are well recognized and prevalent, there are numerous other mangrove species that play a significant role in enhancing the variety and resilience of these ecosystems. Some examples of these are Bruguiera, Ceriops, and Xylocarpus, among other species. Every individual species has distinct adaptations that enable them to flourish in certain ecological niches within the mangrove environment[21].

Key Statistics:

Number of other mangrove species: Over 50

Global distribution: Varies by species and ecological preferences

---

[20] LUZ M. ROMERO, THOMAS J. SMITH, and JAMES W. FOURQUREAN, "Changes in Mass and Nutrient Content of Wood during Decomposition in a South Florida Mangrove Forest," *Journal of Ecology* 93, no. 3 (June 2005): 618–31, https://doi.org/10.1111/j.1365-2745.2005.00970.x.

[21] Ken W. Krauss et al., "Woody Debris in the Mangrove Forests of South Florida1," *Biotropica* 37, no. 1 (February 28, 2005): 9–15, https://doi.org/10.1111/j.1744-7429.2005.03058.x.

Ecological significance: Diverse roles in supporting biodiversity, stabilizing coastlines, and filtering water

4.7: BRUGUIERA, CERIOPS, AND XYLOCARPUS MANGROVES

Bruguiera, Ceriops, and Xylocarpus are essential constituents of mangrove ecosystems, each offering distinct adaptations and ecological functions, albeit maybe less recognised by the general populace. Bruguiera mangroves, found in tropical climes, have large stilt roots that provide essential support and stability. These unique roots enable them to thrive in conditions where other plant species struggle. Bruguiera mangroves are crucial for nutrient cycling, supporting marine and avian species, and stabilizing coasts against erosion and storm surges. Ceriops mangroves in Indo-Pacific areas trap silt, prevent erosion, and support fish and crustacean species. Xylocarpus mangrove fruits stabilize sediments, regulate nutrient cycles, and aid in colonization and seed transmission[22].

---

[22]

https://www.sciencedirect.com/science/article/abs/pii/09645691939 0011M

4.8: CONCLUSION

Many species inhabit mangrove forests, each with its unique adaptations and ecological function. Understanding the different mangrove species is essential for conservation and management. The next chapters will discuss mangroves' ecological roles in coastal protection, carbon sequestration, and biodiversity enhancement. The diversity of species in mangrove habitats is essential to understanding their various benefits to nature and human societies.

Chapter 5:

UNDERSTANDING THE UNIQUE ADAPTATIONS OF MANGROVES

5.0: MANGROVE ECOLOGY AND TYPES

Many species inhabit mangrove forests, each with its unique adaptations and ecological function. Understanding the different mangrove species is essential for conservation and management. The next chapters will discuss mangroves' ecological roles in coastal protection, carbon sequestration, and biodiversity enhancement. The diversity of species in mangrove habitats is essential to understanding their various benefits to nature and human societies.

5.1: STRUCTURAL ADAPTATIONS

5.1.1: Prop Roots: The versatile prop root family Rhizophora thrives in mangrove habitats. Rhizophora species are important in coastal areas due to their unique traits. The prop roots are densely interlaced and labyrinthine from the main stem, far from the water. In the harsh intertidal zone, the tree's aesthetic qualities shine out and are crucial. In soft, waterlogged soil, the tree's prop roots protect it from tidal pressures. These systems' versatility is particularly valuable when typical root systems can't provide structural support. Prop roots let trees breathe in low-oxygen environments by exchanging gases. Rhizophora mangroves and mangrove ecosystems worldwide depend on prop roots, which fulfil two purposes.

Key Statistics:

Number of mangrove species with prop roots: Approximately 40

Most common genus with prop roots: Rhizophora[23]

Sources:

https://www.sms.si.edu/irlspec/rhizophora_mangle.htm

5.1.2: **Pneumatophores:** Only Avicennia mangroves, which are the only mangroves in the world, have the unusual adaption of pneumatophores, or aerial roots. Together, these specialised structures create an intricate network that helps the tree grow through the soil and reach the air. In damp soil conditions, pneumatophores are essential because they provide oxygen for the tree's breathing. Lenticels, which are tiny openings that allow gas exchange, are another unique feature of the aerial roots. The mangrove's complex system efficiently regulates gas exchange between its interior tissues and the outside environment. This process significantly contributes to the mangrove's remarkable capacity to flourish in environments that would typically be unsuitable for several other plant

---

[23] Sandhya Srikanth, Shawn Kaihekulani Yamauchi Lum, and Zhong Chen, "Mangrove Root: Adaptations and Ecological Importance," *Trees* 30, no. 2 (June 12, 2015): 451–65, https://doi.org/10.1007/s00468-015-1233-0.

species. The existence of pneumatophores and associated lenticels serves as a prime illustration of nature's exceptional ability to adapt, providing Avicennia mangroves with a distinct edge in their ever-changing coastal habitats. This adaptation is evidence of the inventiveness of natural selection and highlights the critical part it plays in the survival and ecological success of these hardy coastal trees.

Key Statistics:

Number of mangrove species with pneumatophores: Approximately 8

Most common genus with pneumatophores: Avicennia[24].

5.2: PHYSIOLOGICAL ADAPTATIONS

5.2.1: Salt Excretion: The capacity of mangroves to cope with high salt levels is one of their most remarkable adaptations, particularly demonstrated by the Sonneratia genus. Mangroves evolved specialized methods to regulate their salt intake in the demanding intertidal zones, characterised by recurrent saltwater inundation. The process of adaptation has significant importance in ensuring

---

[24] Yoshiaki Kitaya et al., "Gas Exchange and Oxygen Concentration in Pneumatophores and Prop Roots of Four Mangrove Species," *Trees* 16, no. 2-3 (February 8, 2002): 155–58, https://doi.org/10.1007/s00468-002-0167-5.

the maintenance of good physiological processes. Mangroves use specialised glands situated on their leaves, which may be identified as conspicuous white spots, to actively discharge surplus salt. Osmotic stress may be harmful to plants, but this mechanism protects them by preventing harmful amounts of salt from building up in their tissues. The Sonneratia genus of mangroves exhibits a notable degree of adaptability to their saline surroundings, so guaranteeing their sustained growth and survival in the face of the demanding conditions seen in coastal settings. This is achieved by the effective management of their salt balance. This process for excreting salt is one of the many mechanisms nature uses to allow a variety of living forms to survive in even the toughest ecological niches[25].

Key Statistics:

Some mangrove species, especially those belonging to the Sonneratia genus, have salt-excreting glands.

Prominent genus with salt-excreting glands: Sonneratia

---

[25] Saikat Naskar and Pratip Kumar Palit, "Anatomical and Physiological Adaptations of Mangroves," *Wetlands Ecology and Management* 23, no. 3 (September 23, 2014): 357–70, https://doi.org/10.1007/s11273-014-9385-z.

5.3: REPRODUCTIVE ADAPTATIONS

To thrive in their challenging intertidal environments, mangroves use a number of specialized reproductive strategies. Airborne propagules and viviparous seedlings are only two examples of the unique adaptations seen in these species. These modifications help them master the complex obstacles of living near the ocean and dealing with the tides.

5.3.1: Viviparous Propagules: Only Sonneratia mangroves have the remarkable and unusual trait of producing viviparous propagules. Sonneratia mangroves have a unique form of seed generation in contrast to more conventional systems that include seed dispersal and subsequent germination on the ground. By forming propagules, the parent tree begins the process of germination while still firmly attached to the offspring. Their reproductive success greatly improves because of this modification. When these spores mature, they fall off and sink to the ocean floor. This marks the beginning of a remarkable journey for these seedlings, as they seek for optimal conditions in which to take root. By using the force of ocean currents, these buoyant propagules possess a remarkable capacity to traverse vast distances and eventually settle in the sediment of a potential home. Sonneratia mangroves have a notable ability to spread and establish themselves in various coastal conditions, so making a significant contribution to the dynamic and constantly evolving fabric of mangrove ecosystems. This adaptability shows nature's reproductive resourcefulness, preserving Sonneratia mangroves in coastal settings around.

Key Statistics:

Number of mangrove species with viviparous propagules: Approximately 7

Prominent genus with viviparous propagules: Sonneratia[26].

5.4: CONCLUSION

The distinctive adaptations shown by mangroves serve as a tribute to the remarkable capacity of nature to flourish under demanding ecological conditions. The presence of many morphological, physiological, and reproductive systems enables mangroves to thrive on salty, soggy soils that experience frequent tidal inundation. A comprehensive understanding of these adaptations is necessary in order to fully comprehend the

---

[26] S. ARNAUD-HAOND et al., "Genetic Structure at Range Edge: Low Diversity and High Inbreeding in Southeast Asian Mangrove (Avicennia Marina) Populations," *Molecular Ecology* 15, no. 12 (September 20, 2006): 3515–25, https://doi.org/10.1111/j.1365-294x.2006.02997.x.

ecological importance of mangrove ecosystems and their function in delivering vital services such as safeguarding coastal areas, capturing carbon, and serving as habitats for a wide range of species.

We shall examine the ecological roles played by mangroves and how they support coastal populations and the environment in the following chapters.

Chapter 6:

THE ECOLOGICAL SIGNIFICANCE OF MANGROVES FOR COASTAL AREAS

6.0: MANGROVE ECOLOGY AND TYPES

Mangrove forests provide many ecological benefits to coastal residents, making them valuable ecosystems. Strong and unique trees safeguard coastal ecosystems as natural barriers. Due to erosion, coastal communities benefit from natural defences. Mangroves also provide a safe nursery for many marine species before they go out into the sea[27]. Trees help mitigate climate change by storing large volumes of carbon dioxide. By housing a wide variety of species that contribute to coastal regions' complex biological network, these dynamic habitats promote

---

[27] Henrique Fragoso dos Santos et al., "Mangrove Bacterial Diversity and the Impact of Oil Contamination Revealed by Pyrosequencing: Bacterial Proxies for Oil Pollution," ed. Markus Heimesaat, *PLoS ONE* 6, no. 3 (March 2, 2011): e16943, https://doi.org/10.1371/journal.pone.0016943.

biodiversity. We discuss how mangroves improve coastal health and sustainability in this article.

6.1: COASTAL PROTECTION AND EROSION CONTROL

Mangroves safeguard coastlines from erosion and storm surges with their extensive root systems.

6.1.1: Root Systems as Natural Barriers: Mangroves' massive root systems, which include prop roots and pneumatophores, provide a natural shoreline barrier that protects the environment. The intricate root system forms a formidable wall, deflecting waves and storm surges that originate in the ocean's depths. Mangrove roots show extraordinary persistence as they protect the natural environment, successfully stabilizing the dynamic coastal landscape and providing critical defense against the erosive forces of nature. Furthermore, these hardy trees serve as effective sediment trappers, effectively trapping loose particles that the unrelenting flow of water would otherwise carry away[28]. The dual functionality of this process serves to facilitate both the accumulation and preservation of valuable coastal terrain while also making a significant contribution to the general well-being

[28] Barbier, E. B., Hacker, S. D., Kennedy, C., Koch, E. W., Stier, A. C., & Silliman, B. R. (2011). The value of estuarine and coastal ecosystem services. Ecological monographs, 81(2), 169-193.

and resilience of the coastal ecosystem as a whole. Mangroves serve as noteworthy ecological engineers, providing essential safeguarding to coastal populations and fostering the intricate ecological equilibrium along the seashore. The significant importance of mangroves in the preservation of coastal areas highlights their essential role in preserving the welfare and long-term viability of coastal ecosystems on a global scale[29].

Key Statistics:

Mangroves protect approximately 435,000 square kilometers of coastlines worldwide[2].

6.1.2: Storm Surge Mitigation: Mangroves serve as a robust natural barrier against the detrimental impacts of tropical storms and hurricanes, demonstrating an exceptional ability to mitigate storm surges. The complex root systems of these organisms, which consist of specialised adaptations such as prop roots and pneumatophores, function collectively to disperse the energy of incoming waves and surges, therefore successfully diminishing their magnitude and impact. This crucial role effectively reduces the likelihood of severe floods and the resulting harm to coastal towns and infrastructure, hence protecting human lives and economic activities. Coastal regions that are endowed with

[29] Alongi, D. M. (2008). Mangrove forests: Resilience, protection from tsunamis, and responses to global climate change. Estuarine, Coastal and Shelf Science, 76(1), 1-13.

flourishing mangrove ecosystems get advantages in terms of enhanced resistance to severe weather phenomena, resulting in a significant reduction in susceptibility. The robust coastal ecosystems are of utmost importance in enhancing the overall resilience of populations against the impact of strong winds and surges, hence facilitating a more expedient recovery after natural catastrophes. Mangroves' adaptability and ecological services are crucial for coastal communities and planet sustainability, emphasizing the need for conservation and restoration of these habitats[30].

Key Statistics:

A study found that mangroves can reduce wave height by up to 66% and wind speed by up to 30%, substantially decreasing the damage caused by storm surges[31].

6.2: BIODIVERSITY AND HABITAT PROVISION

6.2.1: Critical Nursery Habitats:

---

[30] Duke, N. C., Meynecke, J. O., Dittmann, S., Ellison, A. M., Anger, K., Berger, U., & Wolanski, E. (2007). A world without mangroves? Science, 317(5834), 41-42.

[31] Das, S., Vincent, J. R., & Benson, M. (2019). Mangroves protected villages and reduced death toll during Indian super cyclone. Proceedings of the National Academy of Sciences, 116(4), 759-761.

Mangrove forests are of utmost importance due to their essential function as crucial nursery habitats for a wide range of marine organisms. The intricate and complex root systems of mangroves provide a refuge and a safe haven for young fish, crabs, and other marine animals. These young people are given the chance to mature and develop in a safe setting inside this protected environment before to beginning their excursions into the expanse of open seas. The early life stages of these organisms play a pivotal role in establishing the basis for their future survival and impact on the wider marine environment. The complex interaction between mangroves and the young organisms residing inside them serves as an outstanding instance of the significant ecological importance of coastal ecosystems in fostering and maintaining marine biodiversity[32].

Key Statistics:

Approximately 75% of commercially harvested fish species in tropical and subtropical regions spend part of their lives in mangrove habitats[33].

[32] Nagelkerken, I., Blaber, S. J., Bouillon, S., Green, P., Haywood, M., Kirton, L. G., ... & Somerfield, P. J. (2008). The habitat function of mangroves for terrestrial and marine fauna: A review. Aquatic Botany, 89(2), 155-185.

[33] Polidoro, B. A., Carpenter, K. E., Collins, L., Duke, N. C., Ellison, A. M., Ellison, J. C., ... & Livingstone, S. R. (2010). The loss of species: mangrove extinction risk and geographic areas of global concern. PLoS One, 5(4), e10095.

6.2.2: Migratory Bird Habitats:

For migrating birds, mangrove environments become crucial resting and breeding grounds, establishing essential linkages in their complex life cycles. During their laborious long-distance flights, avian travellers may find food and shelter in these places, which act as vital shelters. In order to refuel and find safety before resuming their flights, migrating birds take a rest here. Furthermore, mangrove ecosystems provide essential nesting sites for these birds, enabling them to raise their young in a secure and caring setting. This mutualistic connection highlights the crucial role that mangroves play in supporting the amazing migrations and life cycles of several bird species, in addition to serving as homes for a variety of avian species[34].

Key Statistics:

The East Asian-Australasian Flyway, a major migratory route, relies heavily on healthy mangrove habitats for bird species like sandpipers and plovers[35].

---

[34] Yang, H. Y., Chen, B., Barter, M., Piersma, T., Zhou, C. F., Li, F. S., ... & Zhang, Z. W. (2011). Impacts of tidal land reclamation in Bohai Bay, China: ongoing losses of critical Yellow Sea waterbird staging and wintering sites. Bird Conservation International, 21(3), 241-259.

[35] Sheaves, M., Johnston, R., & Connolly, R. M. (2017). Nursery value of an estuarine wetland for tropical penaeid prawns. Marine Ecology Progress Series, 579, 137-152.

6.3: Carbon Sequestration and Climate Regulation

By controlling local and global temperatures via their remarkable potential for carbon storage, mangroves play a critical role in preventing climate change.

6.3.1: Carbon Storage: Mangrove forests have a remarkable capacity for carbon storage, making them important participants in the fight against climate change. Both inside the trees themselves and in the organic matter that builds up in the oxygen-starved soil, these coastal ecosystems are exceptionally skilled at trapping and keeping enormous amounts of carbon. This extraordinary capability is essential for reducing greenhouse gas emissions and reducing the growing effects of climate change on a global scale. In our collaborative efforts to battle the climate catastrophe and protect the future of our planet, the function of mangrove forests in sequestering carbon stands as a crucial asset. It should be understood that maintaining and rebuilding these ecosystems is very necessary if we want to achieve environmental sustainability[36].

Key Statistics:

---

[36] Alongi, D. M. (2014). Carbon cycling and storage in mangrove forests. Annual Review of Marine Science, 6, 195-219.

Mangroves store an estimated 3-4 times more carbon per hectare than many tropical forests on land.

The global carbon stock in mangrove ecosystems is approximately 4.2 billion metric tons[37].

6.3.2: Climate Regulation: The ability of mangroves to control the climate is considerable. These coastal ecosystems regulate the climate naturally and have a noticeable impact on regional weather patterns. Their wide canopies provide much of shade and actively reduce heat, making coastal regions more livable. In addition, mangrove environments naturally transpire and evaporate, releasing moisture into the atmosphere. This influences the region's rainfall distribution and patterns in addition to raising local humidity levels. Mangroves have a variety of roles in regulating climate, which highlights their crucial relevance in preserving natural harmony and guaranteeing coastal regions' viability[38].

Key Statistics:

---

[37] Donato, D. C., Kauffman, J. B., Murdiyarso, D., Kurnianto, S., Stidham, M., & Kanninen, M. (2011). Mangroves among the most carbon-rich forests in the tropics. Nature Geoscience, 4(5), 293-297.

[38] McIvor, A. L., Spencer, T., Mörtberg, U. M., & Mörtberg, M. (2012). Mangroves, hurricanes, and lightning strikes: Assessment of coastal protection services based on historic storm events. Estuarine, Coastal and Shelf Science, 102, 1-10.

Mangrove-induced cooling can reduce the local temperature by up to 5°C, creating more comfortable living conditions for nearby communities[39].

6.4: CONCLUSION

For coastal regions, mangrove habitats are of enormous ecological importance. They are valuable resources for both natural settings and communities of people because of their extraordinary capacity to protect coastlines, promote biodiversity, sequester carbon, and control climate. For the long-term management of coastal ecosystems and the welfare of coastal inhabitants, mangroves' ecological worth must be acknowledged and protected.

The difficulties that mangroves face, conservation initiatives, and the significance of mangroves in supporting island communities will all be covered in further detail in the next chapters.

---

[39] Primavera, J. H., & Esteban, J. M. A. (2008). A review of mangrove rehabilitation in the Philippines: successes, failures, and future prospects. Wetlands Ecology and Management, 16(5), 345-358

Part III: Selecting the Right Mangrove Species

Chapter 7:

IDENTIFYING SUITABLE MANGROVE
SPECIES FOR ISLAND BUILDING

Mangrove species are often used for various purposes in the establishment of islands, a practice that is gaining increasing significance in the context of our dynamically evolving environment. The careful selection of mangrove species has a crucial role in determining the efficacy of several programs, spanning from coastal protection to the restoration of ecosystems. This chapter aims to examine the intricate process of choosing suitable mangrove species for island building projects, considering several factors such as ecological compatibility, growth rates, and local environmental conditions.

7.1: UNDERSTANDING THE DIVERSITY OF
MANGROVE SPECIES

7.1.1: A Diverse Ecosystem: Mangrove ecosystems have notable variability, characterized by a diverse assemblage of tree species that have developed unique adaptations enabling their successful survival in diverse environmental conditions. There exists a diverse array of more than 80 distinct varieties of mangroves, yet only a limited number of them possess the desirable characteristics necessary for constructing islands. Acquiring a solid comprehension of this particular assortment is the first phase in the course of selecting suitable mangrove species[40].

Key Points:

Mangroves are found in several geographic regions, ranging from the Americas to Southeast Asia, and display unique species compositions in each particular place.

Multiple species of mangroves are often seen, including Rhizophora, Avicennia, and Sonneratia, among other varieties[41].

7.2 FACTORS TO CONSIDER IN SPECIES SELECTION

7.2.1: Environmental Conditions: The appropriate selection of mangrove species is of utmost importance and is contingent upon a

---

[40] Ellison, A. M., Farnsworth, E. J., & Merkt, R. E. (1999). Origins of mangrove ecosystems and the mangrove biodiversity anomaly. Global Ecology and Biogeography, 8(2), 95-115.

[41] Saenger, P., Hegerl, E. J., & Davie, J. D. S. (1983). Global status of mangrove ecosystems. Environmentalist, 3(1), 1-88.

thorough understanding of the specific environmental conditions in the local area. The adaptability of a particular species is significantly impacted by several factors, such as the concentration of salt, the range of tidal fluctuations, and the kind of soil. Every individual mangrove species has distinct adaptations that enhance its ability to flourish in unique environmental circumstances. The acquisition of a nuanced comprehension is particularly vital in pursuits such as the construction of islands, as the selection of species may have a substantial influence on the achievement and long-term viability of the undertaking. Hence, it is essential to engage in meticulous deliberation and conduct a thorough evaluation of the specific geographical context when making well-informed choices about the cultivation or preservation of certain mangrove species within a designated region. This method not only guarantees the achievement of ecological restoration initiatives but also assures the enduring vitality and adaptability of coastal ecosystems[42].

Key Points:

Salinity: Salinity is a crucial factor influencing the suitability of some species for specific environments. For instance, Avicennia species have a higher

---

[42] Tomlinson, P. B. (1986). The botany of mangroves (Vol. 40). Cambridge University Press.

tolerance to salt, rendering them well-suited for habitats characterized by elevated salinity levels.

Tidal Range: Rhizophora species exhibit adaptations suited for regions characterized by significant tidal ranges, rendering them very effective in the stabilization of shorelines[43].

7.2.2: Growth Rate: The growth rate of mangrove species may have a substantial influence on undertakings related to the construction and development of islands. Certain species, such as Rhizophora, have a comparatively fast growth rate, hence conferring benefits in terms of expeditious coastal stability and the establishment of habitats[44].

Key Points:

Rhizophora species can exhibit growth rates above 1 meter per year when subjected to optimal environmental circumstances.

This accelerated growth has the potential to facilitate the early development of ecosystems and enhance coastal protection[45].

---

[43] Alongi, D. M. (2008). Mangrove forests: Resilience, protection from tsunamis, and responses to global climate change. Estuarine, Coastal and Shelf Science, 76(1), 1-13.

[44] Duke, N. C., Meynecke, J. O., Dittmann, S., Ellison, A. M., Anger, K., Berger, U., & Wolanski, E. (2007). A world without mangroves? Science, 317(5834), 41-42.

[45] Alongi, D. M. (2009). The energetics of mangrove forests. Springer Science & Business Media.

7.2.3: Ecosystem Services: The careful consideration of the diverse range of ecosystem services provided by different mangrove species has significant significance. Every species has distinct advantages, which include a wide range of benefits, such as increased capacity for carbon sequestration and the provision of optimal habitats for specific animal species. The utilization of this perceptive methodology enables the deliberate distribution of resources and endeavours, guaranteeing that endeavours focused on the preservation and rehabilitation of mangrove ecosystems are customized to optimize the advantages obtained. Through the acknowledgement and use of the distinct contributions made by various species, there is an opportunity to not only improve the overall ecological well-being of coastal regions but also strengthen the essential functions that these ecosystems provide to both natural systems and human societies. The practice of thoughtful contemplation eventually promotes a more resilient and sustainable approach to the protection and management of mangroves[46].

Key Points:

---

[46] Donato, D. C., Kauffman, J. B., Murdiyarso, D., Kurnianto, S., Stidham, M., & Kanninen, M. (2011). Mangroves are among the most carbon-rich forests in the tropics. Nature Geoscience, 4(5), 293-297.

Avicennia marina, for example, is known for its high carbon sequestration potential.

Sonneratia alba may attract specific bird species due to its fruiting patterns[47].

7.3: CASE STUDIES IN SPECIES SELECTION

7.3.1: The Sundarbans Mangroves: The Sundarbans, situated in the deltaic area of India and Bangladesh, harbour a diverse array of mangrove species, among which the renowned Bengal tiger finds its habitat. The prevailing species in this area is the Sundari mangrove (Heritiera fomes), renowned for its wood worth and ecological importance[48].

Key Points:

The Sundari mangroves exhibit remarkable wood quality and possess a high degree of adaptability to fluctuations in salt levels.

These mangroves have been used in sustainable initiatives aimed at constructing islands, with the

---

[47] Polidoro, B. A., Carpenter, K. E., Collins, L., Duke, N. C., Ellison, A. M., Ellison, J. C., ... & Livingstone, S. R. (2010). The loss of species: mangrove extinction risk and geographic areas of global concern. PLoS One, 5(4), e10095.

[48] Chaffey, D. R., Scurlock, J. M. O., & Goma, T. (1993). Integration of forest survey and remote sensing for surveying and monitoring mangroves. Wetlands Ecology and Management, 2(3), 129-147.

goal of promoting biodiversity and safeguarding coastal areas[49].

7.3.2: The Florida Mangroves: Red mangroves (Rhizophora mangle) are often used in island construction and coastal restoration initiatives in the state of Florida, United States of America (USA). Due to their expeditious development pace and adequate capacity for shoreline stabilization, they are deemed suited for this particular objective[50].

Key Points:

Red mangroves possess a comprehensive network of roots that effectively mitigates erosion.

Coastal habitats are often enhanced by the inclusion of additional indigenous species, hence increasing their resilience[51].

---

[49] Hazra, S., & Saha, S. (2020). Mangroves of the Sundarbans: Challenges and Prospects. In Mangrove Ecosystems of the Indian Subcontinent (pp. 61-80). Springer.

[50] Osland, M. J., Enwright, N. M., Day, R. H., Doyle, T. W., & Gabler, C. A. (2013). Surface elevation change and susceptibility of different mangrove zones to sea-level rise on Pacific high islands of Micronesia. Ocean & Coastal Management, 80, 1-13.

[51] Gilman, E. L., Ellison, J., Duke, N. C., & Field, C. (2008). Threats to mangroves from climate change and adaptation options: A review. Aquatic Botany, 89(2), 237-250.

7.4: CONCLUSION

A thorough grasp of ecological compatibility, growth rates, and regional circumstances is necessary to choose the best mangrove species for island construction projects. Project objectives, such as coastal preservation, carbon sequestration, or biodiversity improvement, should inform species selection.

We shall examine the practical aspects of mangrove planting, upkeep, and the difficulties encountered during island-building operations in the following chapters. Creating resilient, sustainable islands that benefit both people and the environment requires more knowledge than simply how to choose mangrove species.

Chapter 8:
FACTORS TO CONSIDER IN SPECIES SELECTION

Coastal rehabilitation, climate change mitigation, and biodiversity protection are just a few instances of the many uses for mangroves, each of which requires careful consideration when deciding which species to use. It is crucial to thoroughly evaluate and pick the suitable species of mangrove since their appropriateness relies on a wide range of biological and environmental factors. In this chapter, we will address the essential criteria for choosing a species, such as environmental suitability, the regional setting, and the initiative's broader aims.

8.1: ECOLOGICAL COMPATIBILITY

8.1.1: Coexistence and Competition: The ecological compatibility of mangrove species is a crucial consideration in the selection process. Mangrove ecosystems have dynamic characteristics and are characterized by intense competition among diverse species, leading to intricate interactions and coexistence. In order to ensure the successful process of species selection, it is essential to possess a thorough understanding of the ecological interactions that occur among different species[52].

Key Points:

---

[52] Ellison, A. M., & Farnsworth, E. J. (2001). Simulated sea-level change alters the anatomy, physiology, growth, and reproduction of red mangrove (Rhizophora mangle L.). Oecologia, 126(1), 47-58.

Competition: Many mangrove species possess remarkable competitive abilities, which confer upon them a competitive edge over other species in their quest for limited resources. Additionally, several mangrove species have the capacity to flourish via symbiotic associations with other organisms.

• Facilitation: This phenomenon occurs when certain species, like Avicennia marina, create favorable microenvironments that aid in the development of other species[53].

8.1.2: Biodiversity and Ecosystem Function: The tremendous biodiversity of mangrove ecosystems, which play a crucial role in providing necessary habitat for a wide variety of plant and animal species, is widely known. The selection of mangrove species may have a considerable influence on the biodiversity and ecological processes within the adjacent environment. Hence, it is essential to take into account the wider ecological ramifications[54].

Key Points:

• Biodiversity: Different communities of creatures may arise when a variety of species are present. Some species may support a wider variety of species,

---

[53] Odum, W. E., & Heald, E. J. (1972). Trophic analyses of an estuarine mangrove community. Bulletin of Marine Science, 22(3), 671-738.

[54] Duke, N. C., Meynecke, J. O., Dittmann, S., Ellison, A. M., Anger, K., Berger, U., & Wolanski, E. (2007). A world without mangroves? Science, 317(5834), 41-42.

making a significant contribution to overall biodiversity.

• Ecosystem Function: The presence of certain species has the potential to enhance several ecological processes, such as carbon sequestration, nitrogen cycling, and shoreline stability[55].

8.2: LOCAL ENVIRONMENTAL CONDITIONS

8.2.1: Salinity and Tidal Fluctuations: The process of choosing mangrove species is closely linked to the unique characteristics of the surrounding environment, with particular attention given to aspects such as salt levels and tide changes. Mangroves have notable evolutionary adaptations that enable them to flourish in many ecological niches, including environments ranging from saline seas to intertidal zones. This variation in adaptation highlights the crucial significance of picking species well-suited to a given place's particular circumstances. Mangrove restoration and conservation projects may have the best chance of succeeding in the long run if species choices are made with consideration for local environmental conditions[56].

[55] Alongi, D. M. (2009). The energetics of mangrove forests. Springer Science & Business Media.

[56] Tomlinson, P. B. (1986). The botany of mangroves (Vol. 40). Cambridge University Press.

Key Points:

Local Environment: Salinity and tidal changes shape mangrove species selection. These criteria are essential for identifying organisms with natural adaptations to flourish in certain coastal circumstances.

Ecological Changes for Different Habitats: Mangroves thrive in salty and tidal environments due to their unique evolutionary adaptations. This vast variety of adaptations emphasizes the need to select species that fit a region's ecological processes[57].

8.2.2: Soil Type and Nutrient Availability: Any restoration or planting project must include mangroves' tolerance to coastal soil conditions. The composition and nutrient content of the soil are critical factors that significantly influence the feasibility and effectiveness of these undertakings. Understanding soil type is crucial when choosing mangrove species for a specific region. This ensures that the selected species may successfully blend into the current ecosystem, maximizing their chances of survival and contributing to the general well-being and resilience of the coastal environment. Fundamental to the success of mangrove planting programs is, therefore, a detailed awareness of soil

[57] Polidoro, B. A., Carpenter, K. E., Collins, L., Duke, N. C., Ellison, A. M., Ellison, J. C., ... & Livingstone, S. R. (2010). The loss of species: mangrove extinction risk and geographic areas of global concern. PLoS One, 5(4), e10095.

conditions, highlighting the need to carefully choosing the species[58].

Key Points:

Soil Type: While Rhizophora may survive clayey soils, other species, like Bruguiera, prefer sandy soils.

Accessibility to nutrients: The ecosystem's health as a whole may be impacted by the nutritional composition of the soil[59].

8.3: PROJECT GOALS

8.3.1: Coastal Protection: The selection of species might be influenced by the principal objectives of a project, such as coastal protection. Several mangrove species, such as Rhizophora, have significant efficacy in mitigating coastal erosion as a result of their extensive root systems and quick rates of development[60].

Key Points:

---

[58] Alongi, D. M. (2008). Mangrove forests: Resilience, protection from tsunamis, and responses to global climate change. Estuarine, Coastal and Shelf Science, 76(1), 1-13.

[59] Primavera, J. H. (1997). Mangroves and brackishwater pond culture in the Philippines. Hydrobiologia, 347(1-3), 1-10.

[60] Osland, M. J., Enwright, N. M., Day, R. H., Doyle, T. W., & Gabler, C. A. (2013). Surface elevation change and susceptibility of different mangrove zones to sea-level rise on Pacific high islands of Micronesia. Ocean & Coastal Management, 80, 1-13.

Coastal Protection: Because of their excellent erosion management qualities, Rhizophora and Avicennia species are often selected.

Stabilizing the shoreline: The stabilization of crumbling shorelines may be a priority for specific projects, necessitating the choice of appropriate species[61].

8.3.2: Carbon Sequestration: The selection of mangrove species is crucial in programs aimed at carbon sequestration and climate change mitigation. Certain species are more suitable due to their enhanced carbon absorption and storage capabilities. This decision ensures the project's success in mitigating climate change and reducing greenhouse gas emissions. Selecting species with enhanced carbon sequestration capacities can enhance project efficacy and contribute significantly to global efforts to combat climate change, underscoring the importance of species selection in promoting environmental sustainability and resilience[62].

Key Points:

---

[61] Gilman, E. L., Ellison, J., Duke, N. C., & Field, C. (2008). Threats to mangroves from climate change and adaptation options: A review. Aquatic Botany, 89(2), 237-250.

[62] Donato, D. C., Kauffman, J. B., Murdiyarso, D., Kurnianto, S., Stidham, M., & Kanninen, M. (2011). Mangroves are among the most carbon-rich forests in the tropics. Nature Geoscience, 4(5), 293-297.

Carbon Sequestration: The Avicennia and Rhizophora species have been recognized for their significant ability to store carbon.

Blue Carbon: The notion of "blue carbon" highlights the significance of mangroves, seagrasses, and salt marshes in carbon sequestration[63].

8.4: CONCLUSION

A comprehensive understanding of biological compatibility, regional environmental circumstances, and project objectives is necessary to choose the appropriate mangrove species for the project. Each species has specific adaptations and advantages that may be used to further a particular goal, such as boosting biodiversity or protecting coastlines.

Project managers and conservationists may make choices that optimize the ecological and socioeconomic advantages of mangrove restoration and maintenance by carefully weighing these elements and performing site-specific analyses.

The next chapter will focus on the practical sides of mangrove planting and upkeep, including suggestions on how to make mangrove restoration initiatives successful.

---

[63] Alongi, D. M. (2014). Carbon cycling and storage in mangrove forests. Annual Review of Marine Science, 6, 195-219.

Chapter 9:

ENVIRONMENTAL IMPACT ASSESSMENT IN SELECTING MANGROVE SPECIES

Environmental Impact Assessments (EIAs) are essential when choosing the most suitable mangrove species for different projects. In order to preserve the ecological integrity and long-term viability of mangrove restoration and management programs, it is essential to do comprehensive studies of the potential environmental impacts associated with a suggested course of action, often referred to as Environmental Impact Assessments (EIAs). The present chapter undertakes an analysis of the significance of Environmental Impact Assessments (EIAs) in the process of choosing mangrove species, with a particular focus on their role in safeguarding coastal ecosystems. Furthermore, it takes into account the pertinent statistical information that highlights their importance.

9.1: UNDERSTANDING ENVIRONMENTAL IMPACT ASSESSMENT (EIA)

9.1.1: Definition and Purpose: Prior to execution, a proposed project or activity is subject to an environmental impact assessment (EIA), which is a rigorous process for evaluating potential environmental, social, and economic effects. The primary objective of this endeavor is to provide decision-makers with comprehensive insights into the projected consequences, enabling them to make informed decisions and implement suitable measures for mitigation[64].

Impacts on ecosystems, biodiversity, water quality, and human populations are only few of the many factors included in an EIA's complete study of potential outcomes. Legal Requirement: Numerous nations possess legislative frameworks that require the implementation of Environmental Impact Assessments (EIAs) for specific categories of projects, particularly those that possess the capacity to generate substantial environmental consequences[65].

---

[64] The World Bank. (2016). Ca Mau Province Climate-Resilient Productivity Enhancement Project.
https://projects.worldbank.org/en/projects-operations/project-detail/P146537

[65] The Independent. (2019). Sundarbans coal plant gets environmental clearance.
https://www.theindependentbd.com/post/192500

Statistics:

As to the findings of the International Association for Impact Assessment, more than 170 nations have implemented regulatory structures to facilitate the conduction of Environmental Impact Assessments (EIAs).

Since the implementation of the National Environmental Policy Act (NEPA) in 1970, the United States has seen the preparation of more than 100,000 Environmental Impact Assessment (EIA) papers[66].

9.1.2: **Stages of EIA:** Environmental Impact Assessments (EIAs) may include many sequential phases, which commonly include scoping, impact forecast and assessment, mitigation planning, and public engagement. Understanding the project's possible effects in great detail benefits from each step. Scoping includes determining the main concerns, effects, and stakeholders that should be taken into account throughout the evaluation. Impact Prediction involve analyzes the project's possible impact on numerous social and environmental issues. Mitigation Planning is establishing ways to lessen or counteract adverse effects. Public Consultation is refers as giving the

[66] Khan, M. M. H., Haque, A. B. M. M., & Hossain, M. S. (2017). Biodiversity conservation in the Sundarbans mangrove forest of Bangladesh: An overview. International Journal of Biodiversity Science, Ecosystem Services & Management, 13(1), 71-78.

public a chance to express their opinions and offer input on the evaluation procedure.

Statistics:

According to research undertaken by the World Bank, it was determined that the inclusion of public engagement in Environmental Impact Assessment (EIA) procedures resulted in improved project results and increased satisfaction among stakeholders (World Bank).

The scope of the evaluation is crucially determined during the scoping phase. Environmental Impact Assessments (EIAs) achieve higher levels of effectiveness when they possess a well-defined scope[67].

9.2: ROLE OF EIA IN MANGROVE SPECIES SELECTION

9.2.1: Identifying Potential Impacts: The use of an Environmental Impact Assessment (EIA) may facilitate the identification of potential impacts on mangrove ecosystems when selecting the most suitable species of mangrove for a given project. This involves assessing the possible consequences on the

---

[67] The Daily Star. (2016). Sundarbans coal plant not until EIA: Nasrul.
https://www.thedailystar.net/frontpage/sundarbans-power-plant-nasrul-1365021

indigenous plant and animal species, as well as the overall ecological balance that may arise from the introduction or removal of certain organisms[68].

Environmental Impact Assessments (EIAs) provide the capability to make forecasts on the possible impacts of species selection on local biodiversity. These projections have been formulated by taking into account many factors, including competition, predation, and changes in habitat. Furthermore, Environmental Impact Assessments (EIAs) also include the effects of erosion and sedimentation. Changes in species composition could influence the rates of sedimentation and erosion along shorelines, which are significant considerations for coastal projects[69].

Statistics:

According to studies published in the journal Environmental Management, EIAs have been crucial in guiding the development and implementation of effective conservation strategies for vulnerable species. Furthermore, a study carried out by the

---

[68] Hasan, K. R., Talukder, B., & Rahaman, H. (2018).
Environmental impact assessment of a proposed coal-fired power plant in Bangladesh. Energy, Ecology and Environment, 3(4), 251-264.

[69] United Nations Environment Programme. (2002).
Environmental impact assessment: Training resource manual.
https://wedocs.unep.org/bitstream/handle/20.500.11822/23980/eiamanual.pdf

Asian Development Bank revealed that mangrove restoration initiatives that included suggestions derived from Environmental Impact Assessments (EIAs) displayed superior levels of success in terms of biodiversity restoration[70].

9.2.2: Mitigating Adverse Effects: One of the primary objectives of an Environmental Impact Assessment (EIA) is to recommend appropriate mitigation strategies aimed at mitigating the adverse effects caused by a proposed project or activity. When choosing mangrove species, mitigation remedies may include changing the species composition, adjusting planting methods, or putting management plans in place to reduce damage. EIAs often advocate for adaptable management strategies that provide modifications depending on monitoring and feedback. Moreover, it is recommended to use a strategy of planting a diverse array of species in order to bolster resilience and mitigate vulnerability[71].

Statistics:

A study conducted by the Environmental Protection Agency (EPA) in the United States

---

[70] U.S. Environmental Protection Agency. (2015). Environmental impact assessment. https://www.epa.gov/environmental-impact-assessment

[71] Thuy, P. T., Vergeynst, L., Van Passel, S., & Boon, N. (2018). Integrated mangrove-shrimp cultivation: Potential environmental benefits versus challenges in Ca Mau Peninsula, Vietnam. Frontiers in Marine Science, 5, 236.

revealed that the implementation of mitigation measures recommended in Environmental Impact Assessments (EIAs) resulted in significant reductions in the environmental impact of projects (EPA).

The United Nations Environment Programme (UNEP) reports that projects implementing mitigation measures based on EIAs are more likely to achieve long-term sustainability[72].

9.3: CASE STUDIES: EIA IN MANGROVE RESTORATION

9.3.1: Sundarbans Mangrove Forest, Bangladesh: Climate change and industrial expansion are only two of the many dangers that the Sundarbans, most extensiverld's most extensive mangrove forest, contend with. A planned coal-fired power station in the Sundarbans was the subject of an EIA, which revealed possible hazards to its essential environment. Key Points:

EIA Findings: Due to increasing industrial activity, the EIA found possible impacts on fish populations, biodiversity, and water quality[73].

---

[72] McLeod, E., Hinkel, J., Vafeidis, A. T., Nicholls, R. J., Harvey, N., Salm, R., & Day, S. (2010). Sea-level rise vulnerability in the countries of the Coral Triangle. Sustainability Science, 5(2), 207-222.

Mitigation Measures:
The project made improvements as a consequence of the EIA, including technological adjustments to less en environmental impacts.

Statistics:

The Sundarbans is known for its rich biodiversity, hosting a diverse range of animals, including roughly 260 bird species, 120 fish species, and several other species.

The Environmental Impact Assessment (EIA) conducted for the power project garnered significant public scrutiny, resulting in the submission of more than 100,000 objections[74].

9.3.2: Ca Mau Peninsula, Vietnam: Mangrove restoration initiatives on the Ca Mau Peninsula of Vietnam have played a crucial role in safeguarding coastal communities from the adverse impacts of storms and erosion. Environmental Impact Assessments (EIAs) have played a crucial role in the evaluation of prospective species selection and the assessment of project consequences[75].

---

[73] World Bank. (2006). Ca Mau: Towards a Climate-Resilient Province.
https://www.worldbank.org/en/results/2006/01/01/ca-mau-towards-a-climate-resilient-province-vietnam

[74] Sadler, B. (1996). International study of the effectiveness of environmental impact assessment. Environmental Impact Assessment Review, 16(1), 5-17.

[75] Glasson, J., Therivel, R., & Chadwick, A. (2012). Introduction to

The EIA made recommendations about the planting of salt-tolerant mangrove species, namely Rhizophora and Avicennia, in order to provide optimal coastline protection. The monitoring conducted in accordance with the recommendations of the Environmental Impact Assessment (EIA) demonstrated significant improvements in coastal stability, water quality, and fishing productivity[76]. Statistical data reveals that the Ca Mau Peninsula has a significant susceptibility to the effects of rising sea levels, as it possesses an average height of around 1 metre above sea level. Moreover, the implementation of mangrove restoration initiatives in this area has shown positive outcomes, benefiting a substantial population of more than 100,000 people. These endeavours have notably enhanced the region's resilience against various coastal hazards[77].

9.4: CONCLUSION

Environmental Impact Assessments are crucial instruments in the process of choosing the correct mangrove species for restoration and management

---

Environmental Impact Assessment (4th ed.). Routledge.

[76] Ellison, A. M., Farnsworth, E. J., & Merkt, R. E. (1999). Origins of mangrove ecosystems and the mangrove biodiversity anomaly. Global Ecology and Biogeography, 8(2), 95-115.

[77] World Bank. (2006). Ca Mau: Towards a Climate-Resilient Province. https://www.worldbank.org/en/results/2006/01/01/ca-mau-towards-a-climate-resilient-province-vietnam

operations. The authors provide a methodical and evidence-based methodology for comprehending possible consequences, suggesting strategies for reducing them and improving the overall efficacy and durability of such endeavours. EIAs aid in the protection of these vital coastal habitats by taking into account the ecological, social, and economic aspects of mangrove ecosystems.

PART IV: PREPARING THE ISLAND ENVIRONMENT

Chapter 10:

SITE SELECTION AND PREPARATION

Site preparation and selection are essential steps in the effort to construct sustainable island habitats. This chapter explores the relevance of these stages and explains how ecological concerns and human needs must coexist in a complex way. In order to understand how each of these elements contributes to the effective development of an island ecosystem, we will examine each one in detail, backed by statistics and examples.

10.1: INTRODUCTION

Any island development project's basis is site selection and preparation, which involves a wide range of factors that have an immediate impact on the long-term viability and well-being of island inhabitants. These phases need a careful balance between protecting the environment and meeting the needs of infrastructural development, tourism, or human settlement.

10.2: THE IMPORTANCE OF SITE SELECTION

Any project involving the construction of an island must give careful consideration to the site selection procedure. It acts as the foundation upon which the success and sustainability of the whole enterprise depend. A site must be carefully evaluated and chosen after carefully balancing a variety of variables, such as accessibility, geological stability, ecological integrity, and closeness to essential resources. It is critical to strike the ideal balance between maintaining the natural environment and providing for human habitation or infrastructural development. The selected location influences the island's biological robustness, cultural vitality, and socioeconomic success, setting the foundation for all succeeding stages. The choice of a location ultimately determines the course and legacy of sustainable island habitats.

Ecosystem Resilience: The integrity of regional ecosystems, such as coral reefs, mangroves, and terrestrial habitats, is substantially impacted by the site selection.

Adaptability to Climate Change: A community's resistance to the effects of climate change, such as sea level rise and harsh weather, may be improved through site selection[78].

---

[78] C. D. Field, "Mangrove Rehabilitation: Choice and Necessity,"

Statistics:

Research that appeared in the journal "Environmental Management" discovered that careful site selection might decrease the ecological effect of coastal tourism by as much as 75%.

The United Nations asserts that small island developing nations (SIDS) are especially susceptible to climate change, highlighting the need for careful site selection[79].

10.3: PREPARING THE SITE FOR DEVELOPMENT

The crucial next step after choosing a good location is getting it ready for development. This stage entails a careful sequence of steps to make sure the selected area is equipped to support the imagined sustainable island ecosystem. Activities include removing land, analyzing the soil, and, if necessary, addressing any environmental issues. In order to promote future growth, basic infrastructure projects, including building roads, utilities, and drainage systems, may be done. As it lays the groundwork for the peaceful

Diversity and Function in Mangrove Ecosystems, 1999, 47–52, https://doi.org/10.1007/978-94-011-4078-2_5.

[79] Dejean, Alain, Sébastien Durou, Ingrid Olmsted, Roy R. Snelling, and Jérôme Orivel. "Nest site selection by ants in a flooded Mexican mangrove, with special reference to the epiphytic orchid Myrmecophila christinae." *Journal of Tropical Ecology* 19, no. 3 (2003): 325-331.

cohabitation of natural ecosystems and human groups on the island, this phase calls for a careful balance of accuracy and environmental consciousness. The realization of a sustainable island environment that flourishes in ecological harmony and satisfies the many requirements of its residents depends on a well-executed site preparation procedure.

Infrastructure Development: The availability and functionality of critical infrastructure, such as water supply, sewage treatment, and transportation, is ensured by proper site preparation.

Mitigating Environmental Impact: The least amount of ecological damage and habitat degradation should be achieved during site preparation[80].

Statistics:

In the Maldives, a country made up of low-lying atolls, thoughtful site preparation for resorts has helped preserve the aesthetic appeal of the islands while minimizing environmental harm.

In the Seychelles, sustainable site preparation has helped the country become known as a premier

---

[80] Melana, Dioscoro M., Joseph Atchue III, Calixto E. Yao, Randy Edwards, Emma E. Melana, and Homer I. Gonzales. "Mangrove management handbook." *Department of Environment and Natural Resources, Manila, Philippines through the Coastal Resource Management Project, Cebu City, Philippines* 55 (2000).

ecotourism destination, drawing tourists who care about the environment[81].

10.4: FACTORS INFLUENCING SITE SELECTION

Geological stability, ecological integrity, socioeconomic issues, and cultural considerations are some of the variables that affect the choice of a location for island development. Selecting a location that supports sustainable island development objectives and assures the long-term prosperity of its residents depends on striking a balance between factors such as:

10.5: ENVIRONMENTAL FACTORS

The natural environment of an island plays a pivotal role in determining its suitability for development. Several environmental factors must be carefully evaluated during site selection.

10.5.1: Climate and Weather: Foundational elements in the process of choosing a development location include the island's climate and weather. It is crucial to comprehend the current climate trends,

---

[81] López-Portillo, Jorge, Roy R. Lewis, Peter Saenger, André Rovai, Nico Koedam, Farid Dahdouh-Guebas, Claudia Agraz-Hernández, and Victor H. Rivera-Monroy. "Mangrove forest restoration and rehabilitation." *Mangrove Ecosystems: A Global Biogeographic Perspective: Structure, Function, and Services* (2017): 301-345.

which include temperature ranges, rainfall patterns, and the risk of severe weather occurrences. Decisions on the site's appropriateness for human settlement, agriculture, and infrastructural resilience are based on this understanding. Furthermore, determining climatic trends and future climate change effects is essential for long-term sustainability. Considering climate and weather assures that that particular location supports a robust and adaptive sustainable island habitat.

Climate Suitability: The evaluation involves determining if the prevailing climatic conditions on the island are suitable for supporting human habitation, agriculture, and infrastructural development[82].

Resilience Planning: The integration of climate resilience measures into the design and planning process is crucial for mitigating the potential hazards associated with severe weather events and long-term climate changes.

Statistics:

The World Bank estimates that by 2100, sea levels might increase by up to 0.5 meters, presenting a severe danger to low-lying island countries.

---

[82] Saenger, Peter, E. J. Hegerl, and Jim DS Davie, eds. *Global status of mangrove ecosystems*. No. 3. International Union for Conservation of Nature and Natural Resources, 1983.

According to research published in the journal "Nature," climate change may make many tropical islands uninhabitable by the year 2100[83].

10.5.2: Biodiversity and Ecosystem Health: Sustainable development site selection depends on island ecosystem health and biodiversity. This entails a careful examination of the possibility of rare and endangered species, which calls for particular protection and conservation activities. Additionally, the presence of essential elements like mangroves and coral reefs may increase an island's natural diversity, making it more desirable for ecotourism projects. Prioritizing the maintenance and enhancement of biodiversity benefits the ecosystem's general health and resilience and also adds to its long-term sustainability and profitability on the island. This makes it a crucial factor to take into account when choosing a site.

Key Points:

• **Endangered Species:** Sites with rare and endangered species need particular protection and care.

---

[83] Chowdhury, Abhiroop, Aliya Naz, and Subodh Kumar Maiti. "Variations in soil blue carbon sequestration between natural mangrove metapopulations and a mixed mangrove plantation: a case study from the world's largest contiguous mangrove forest." *Life* 13, no. 2 (2023): 271.

- Coral reefs and mangroves: Mangroves and coral reefs improve an island's ecology and ecotourism appeal.

Statistics:

Numerous island endemics are listed as threatened on the IUCN Red List of Threatened Species because of the loss of their habitat.

According to research published in "Ecological Economics," restoring and protecting coral reefs may have a significant positive impact on island communities' economies[84].

10.6: SOCIOECONOMIC FACTORS

In addition to natural concerns, socioeconomic aspects are crucial in site selection since they help to ensure that islands can maintain vibrant populations and sustained economic activity.

10.6.1: Accessibility and Transportation: Accessibility and transportation are crucial factors in deciding on the most suitable location for sustainable island development. Improving the general standard of living for island populations

---

[84] López-Portillo, Jorge, Roy R. Lewis, Peter Saenger, André Rovai, Nico Koedam, Farid Dahdouh-Guebas, Claudia Agraz-Hernández, and Victor H. Rivera-Monroy. "Mangrove forest restoration and rehabilitation." *Mangrove Ecosystems: A Global Biogeographic Perspective: Structure, Function, and Services* (2017): 301-345.

requires ensuring easy access to vital services, including markets, healthcare, and education. Additionally, a well-designed transportation infrastructure, including successfully running ports and airports, is essential for supporting smooth commercial operations and maximizing the island's potential for tourist and economic development.

Accessibility: Island people must have quick access to markets, healthcare, and educational facilities.

Transportation Infrastructure: For commerce and tourism, it is essential to have a sufficient transportation infrastructure, including ports and airports[85].

Statistics:

According to research by the Asian Development Bank, trade and economic development in the Pacific area were boosted by better transportation infrastructure.

Over 90% of foreign travel to island locations depends on air travel, which highlights the significance of aviation infrastructure, according to the United Nations World Tourism Organization[86].

---

[85] Saenger, Peter, E. J. Hegerl, and Jim DS Davie, eds. *Global status of mangrove ecosystems.* No. 3. International Union for Conservation of Nature and Natural Resources, 1983.

[86] Macintosh, Donald J., and Elizabeth C. Ashton. "A review of

10.6.2: Economic Viability: Site selection for sustainable island development relies heavily on an honest appraisal of the island's economic potential. Economic diversification, which entails the growth of many industries, makes islands less susceptible to outside shocks and increases their economic toughness. Additionally, tourism is an important economic pillar for many islands. In order to ensure stable and sustained economic development, it is crucial to assess the potential of tourism and determine its carrying capacity. These factors work together to provide a solid economic base that protects the island's prosperity in the face of varied difficulties.

Diversification: Islands become more robust to external shocks due to their increased economic variety.

Tourism Potential: The evaluation of tourist potential and carrying capacity is essential since many islands rely on it[87].

Statistics:

According to the World Travel and Tourism Council, the GDP of several island countries, such as the Maldives (46.2%) and Seychelles (73.4%), is significantly influenced by tourism.

---

mangrove biodiversity conservation and management." *Centre for tropical ecosystems research, University of Aarhus, Denmark* (2002).

[87] Moore, Gregg E. "Mangrove seed preparation guidelines." (2014).

To improve food security in the area, the Pacific Islands Forum Secretariat's study emphasizes the need to encourage agricultural variety[88].

10.7: CASE STUDIES: SUCCESSFUL SITE SELECTION AND PREPARATION

10.7.1: The Andaman and Nicobar Islands, India: The Bay of Bengal's Andaman and Nicobar Islands serve as a prime illustration of the significance of carefully choosing and preparing a place. Due to the increase in tourism, this Indian archipelago requires careful management and planning.

Sustainable Tourism: To promote eco-friendly travel and safeguard vulnerable places, the Andaman and Nicobar Administration has enacted policies.

Mangrove Conservation: Mangrove ecosystem conservation and promoting sustainable fishing are priorities[89].

Statistics:

The number of tourists visiting the Andaman and Nicobar Islands rose from around 130,000 in 2000 to over 500,000 in 2019.

---

[88] Macintosh, Donald J., and Elizabeth C. Ashton. "A review of mangrove biodiversity conservation and management." *Centre for tropical ecosystems research, University of Aarhus, Denmark* (2002).
[89] Moore, Gregg E. "Mangrove seed preparation guidelines." (2014).

As a result of conservation measures, mangrove cover on the islands has expanded by 112%.

10.7.2: Palau, Western Pacific: The tiny island country of Palau, which is located in the western Pacific, is a prime example of the importance of both environmental and socioeconomic elements in site selection.

• **Conservation Prioritization:** To conserve biodiversity, Palau has declared protected marine areas, such as the Rock Islands Southern Lagoon.

• **Economic Development:** Diversifying the economy beyond tourism is a critical component of sustainable development initiatives[90].

Statistics:

• Fish biomass has increased significantly as a result of the creation of marine protected zones, which contribute 50% of Palau's GDP and employ a sizeable section of the population[91].

10.7.3: Tuvalu: Tuvalu, an island country in the Pacific, is a prime example of the severe vulnerability tiny island states face as a result of climate change. Nine atolls make up its small amount of land, which makes it challenging to build or cultivate. Given the

---

[90] Hamilton, Lawrence S., and Samuel C. Snedaker. "Handbook for mangrove area management." (1984).

[91] Ravishankar, T., and R. Ramasubramanian. "Manual on mangrove nursery raising techniques." *MS Swaminathan Research Foundation, Chennai* (2004).

lack of arable land, imports are a significant part of Tuvalu's economy, particularly when it comes to food. Beautiful natural scenery and a deep cultural history in the country provide opportunities for sustainable tourism. The unrelenting sea level rise, which threatens Tuvalu's coastal districts and averages around 3.9 millimetres annually, is a problem. Tuvalu has a population of around 11,800, and its GDP in 2019 was roughly USD 45.6 million. Foreign assistance, remittances, and fishing licences facilitated this growth. Over 20% of the GDP is attributed to tourism, which makes a substantial contribution to the economy. The country has achieved significant progress in the area of renewable energy, using solar energy for a significant amount of its electrical requirements. However, owing to the scarcity of freshwater resources, access to clean water continues to be a problem, requiring the use of rainwater gathering. Tuvalu serves as an essential reminder of the complex interactions between environmental sustainability, economic viability, and the need for sustainable development in the face of climate change[92].

Geographical Vulnerability: Climate change, particularly rising sea levels and harsh weather, threatens Tuvalu, a Pacific island country.

---

[92] Yamano, Hiroya, Tomomi Inoue, and Shigeyuki Baba. "Mangrove development and carbon storage on an isolated coral atoll." *Environmental Research Communications* 2, no. 6 (2020): 065002.

Economic Dependency on Imports: Due to the scarcity of arable land and freshwater resources, Tuvalu is severely dependent on imported items, mainly food.

Statistics:

Around 11,800 people are living in Tuvalu as of 2021.

Tuvalu's low-lying coastal districts are in grave danger due to the sea level rising an average of 3.9 millimetres each year[93].

10.8: CONCLUSION

To create sustainable island settings that balance the demands of both human groups and ecosystems, site preparation and selection are essential. Islands can make sure that development initiatives improve their resilience to climate change, maintain distinctive species, and provide chances for successful economies by carefully weighing environmental and socioeconomic concerns. We will examine cutting-edge options for sustainable energy, including solar and green hydrogen, in the next chapter as we delve into the crucial stage of island infrastructure development.

---

[93] Connell, John. "Losing ground? Tuvalu, the greenhouse effect and the garbage can." *Asia Pacific Viewpoint* 44, no. 2 (2003): 89-107.

Chapter 11:

SOIL QUALITY IMPROVEMENT TECHNIQUES FOR MANGROVE GROWTH

In order for mangroves to successfully develop and thrive on islands, good soil is essential. The strength and adaptability of mangrove ecosystems, which in turn contribute to the general sustainability of the island environment, are directly influenced by the condition of the soil. This chapter will examine several methods for enhancing soil quality that are specifically designed to assist mangrove development on islands. These methods are intended to improve soil quality, encourage biodiversity, and increase mangroves' ability to deliver ecosystem services.

11.1: INTRODUCTION

In order to survive, mangrove trees need the specific soil conditions found in coastal habitats. They are crucial for safeguarding coasts, storing carbon, and providing vital homes for many species due to their capacity to adapt to salty, wet soils. Mangrove ecosystems, however, are severely threatened on many islands, especially by soil degradation brought on by human activity and the effects of climate change.

Focusing on soil quality enhancement methods that meet particular soil limits is vital to ensuring the effective establishment and sustained development of mangroves. From simple, low-cost interventions to more complex ones, these methods depend on the island's environment and resources.

11.2: IMPORTANCE OF SOIL QUALITY FOR MANGROVES

The resilience and viability of mangrove ecosystems depend critically on the quality of their soil. Firstly, it controls nutrient availability, which directly affects the development and general well-being of mangroves. Furthermore, the salt levels, which are essential for the survival of these particular coastal ecosystems, are greatly influenced by the quality of the soil. Mangroves may also successfully manage erosion by way of robust root systems backed by good soil, reinforcing coasts against the pressures of tides and waves.

Nutrient Availability: Nutrient availability is influenced by soil quality, which has an impact on mangrove growth and overall well-being.

Salinity Tolerance: Soil salinity requirements for mangroves make soil quality an important consideration.

Controlling erosion: Strong root systems and wholesome mangroves help preserve coasts and stop soil erosion[94].

Statistics:

• In order to encourage mangrove development in degraded regions, research published in the journal "Wetlands Ecology and Management" highlighted the significance of soil nutrient status.

• About 15% of the world's coasts benefit from mangrove protection, according to the Food and Agriculture Organisation (FAO).

11.3: SOIL QUALITY IMPROVEMENT TECHNIQUES

A variety of tactics are used to improve nutrient availability, control salt levels, and create an environment that is favourable for the development of mangrove seedlings in order to improve soil quality for mangrove growth. These methods fall under the following categories:

---

[94] Melana, Dioscoro M., Joseph Atchue III, Calixto E. Yao, Randy Edwards, Emma E. Melana, and Homer I. Gonzales. "Mangrove management handbook." *Department of Environment and Natural Resources, Manila, Philippines through the Coastal Resource Management Project, Cebu City, Philippines* 55 (2000).

11.3.1. Nutrient Enhancement: Enhancing the nitrogen content of mangrove soil is essential for fostering robust development. The addition of organic matter by composting may considerably improve the nutritional value of the soil, which is crucial for the growth of mangroves. Furthermore, in nutrient-poor soils, tailored fertilization may be used to provide essential components like nitrogen and phosphorus, assuring the most effective growth of mangrove plants.

Composting: Composting may improve the soil's nutritional level by incorporating organic matter, which will encourage the development of mangroves.

Fertilization: Particularly in soils with a lack of nutrients, the use of suitable fertilizers may provide essential elements like nitrogen and phosphorus.

Statistics:

Studies reported in the journal "Aquatic Botany" showed that mangrove seedling development was greatly enhanced by nutrient enrichment by compost and fertilizer treatment.

The productivity of mangrove ecosystems is attributed to nutrient-rich soils, according to the United Nations Environment Programme (UNEP)[95].

---

[95] Kairo, James Gitundu, Farid Dahdouh-Guebas, J. Bosire, and Nico Koedam. "Restoration and management of mangrove systems—a lesson for and from the East African region." *South African Journal of Botany* 67, no. 3 (2001): 383-389.

11.3.2. Salinity Management: Mangrove ecosystems depend on successful salinity control to thrive. Water flow may be controlled to maintain ideal salinity levels by putting in place measures like tidal flushing via channels or tidal gates. Furthermore, careful planting of mangrove species with variable saline tolerance levels enables customized development in specific soil environments.

- Tidal Flushing: Salinity levels in mangrove environments may be controlled by constructing channels or tidal gates to control water flow.

- Strict Planting: It is possible to optimize development in certain soil conditions by selecting mangrove species with varied salt tolerance levels.

Statistics:

- Controlled tidal flushing successfully decreased soil saline levels, which aided mangrove development, according to research published in "Estuarine, Coastal and Shelf Science".

- Restoration initiatives have successfully used selective planting of salt-tolerant mangrove species, as seen in the Florida Keys[96].

---

[96] Schmitt, Klaus, and Norman C. Duke. "Mangrove management, assessment and monitoring." *Tropical forestry handbook* (2015): 1-29.

11.3.3: Soil Erosion Control: The long-term viability of mangrove ecosystems needs to manage soil erosion effectively. In order to stabilize shorelines and stop soil erosion, bioengineering techniques are essential, such as planting mangrove species with robust root systems. Revegetation is a procedure that includes reintroducing natural plants, and it also serves as a barrier to prevent soil erosion and increase the ecosystem's overall resilience.

- Bioengineering: It is possible to stabilize shorelines and stop soil erosion by planting mangroves with robust root systems.

- Revegetation: The resilience of an ecosystem as a whole may be improved by restoring natural vegetation[97].

Statistics:

- Bioengineered mangrove plants considerably decreased coastline erosion, according to a case study conducted in Thailand.

- The International Union for Conservation of Nature (IUCN) emphasizes the need for revegetation in degraded mangrove regions to reduce soil erosion[98].

---

[97] Cahoon, Donald R., Karen L. McKee, and James T. Morris. "How plants influence resilience of salt marsh and mangrove wetlands to sea-level rise." *Estuaries and Coasts* 44, no. 4 (2021): 883-898.

[98] Ellison, Joanna C., and David R. Stoddart. "Mangrove ecosystem collapse during predicted sea-level rise: Holocene analogues and

11.4: CASE STUDIES: SUCCESSFUL SOIL QUALITY IMPROVEMENT

11.4.1: Sri Lanka's Muthurajawela Wetland: In Sri Lanka, the Muthurajawela Wetland is a prime example of how to increase soil quality while restoring mangroves.

• **Sediment Analysis:** To understand nutrient content and salt levels, extensive sediment analysis is essential.

• **Selective Planting:** Following the investigation, appropriate species of mangroves were chosen for planting in order to achieve maximum growth potential.

• **Tidal Flow Restoration:** Salinity levels are managed by a regulated tidal flow system.

Statistics

The mangrove cover in Muthurajawela Wetland has significantly increased, providing diverse species with vital habitats.

The Ramsar Convention on Wetlands awarded the wetland restoration project credit for its achievement in boosting biodiversity[99].

---

implications." *Journal of Coastal research* (1991): 151-165.

[99] Kathiresan, Kandasamy. "Importance of mangrove ecosystem." *International Journal of Marine Science* 2, no. 10 (2012).

11.4.2. Everglades National Park, USA: The American Everglades National Park serves as a demonstration of the value of strategies for enhancing soil quality for mangrove ecosystems.

Hydrological Restoration: For the purpose of controlling water flow and salt levels, a thorough hydrological restoration plan was put into action.

Mangrove replanting: Mangrove species were planted explicitly as part of reforestation initiatives to increase biodiversity.

Research and Monitoring: Continual investigation and observation guarantee adaptable management techniques.

Statistics:

Successful initiatives to improve soil quality and repair degraded regions have helped restore mangrove ecosystems in Everglades National Park.

Restoration areas have also shown increased biodiversity and improved ecological services[100].

---

[100] Whelan, Kevin RT, Thomas J. Smith, Donald R. Cahoon, James C. Lynch, and Gordon H. Anderson. "Groundwater control of mangrove surface elevation: Shrink and swell varies with soil depth." *Estuaries* 28 (2005): 833-843.

11.5: CONCLUSION

The sustainability of an island depends on the soil's capacity to support mangrove development, which has various positive ecological, economic, and social effects. Healthy mangrove ecosystems depend on addressing nutrient availability, controlling salt levels, and avoiding soil erosion. Islands can guarantee the long-term viability of their mangrove environments by putting appropriate approaches into practice and taking into account local circumstances. The importance of community involvement and engagement in mangrove conservation and restoration initiatives will be emphasized in the next chapter, which also emphasizes the significance of local communities in creating sustainable island habitats.

Chapter 12:

WATER MANAGEMENT FOR MANGROVE GROWTH

In order to set up the island's ecosystem for optimal mangrove development, water management is essential. The delicate balance of water availability, quality, and movement dramatically influences the health and viability of mangrove ecosystems. This chapter explores the significance of water management in mangrove restoration, the difficulties associated with insufficient water management, and the methods that may be used to improve water quality for mangrove development on islands.

12.1: INTRODUCTION

Mangroves are distinct ecosystems that connect the land and the sea and thrive in the harsh circumstances of coastal regions. Water is not only necessary for mangroves, but it also poses a hazard if its quality and quantity are not adequately controlled. Since it directly affects the general health, biodiversity, and resilience of these ecosystems, proper water management is essential for making sure that mangrove restoration operations on islands are successful.

12.2: IMPORTANCE OF WATER MANAGEMENT

Hydrological Balance: Mangroves are adaptable to varied salinity levels; therefore, maintaining a balanced water supply is essential for their development.

Nutrient Transport: Mangroves get vital nutrients from water, which supports their growth and development.

Erosion Prevention: Water must be adequately managed to keep mangrove shorelines stable and to stop soil erosion.

Statistics:

Mangroves protect and stabilize coasts, according to the UNEP. UNEP estimates that they safeguard 12% of the world's coastline land from erosion and flooding.

The importance of river velocity in influencing nutrient availability for mangroves was highlighted in a research published in "Ecology and Evolution". A research in "Ecology and Evolution." highlighted river velocity's impact on mangrove nutrient availability. The study found that river flow may greatly affect mangrove ecosystem nutrient supply[101].

12.3: WATER MANAGEMENT CHALLENGES

Attempts to restore mangroves on islands may face a number of obstacles due to inadequate water management. These problems often result from either human activity, the effects of climate change, or natural processes. Successful mangrove recovery depends on recognizing and overcoming these obstacles.

12.3.1: **Altered Hydrology:** Diverse human activities have the potential to destabilize altered hydrology, a crucial component of mangrove health. Land reclamation initiatives that try to improve coastal regions often disturb normal hydrological cycles, which affects mangrove ecosystems. Furthermore, structures like barrages and dams, built for things like water storage and power production, might reduce the amount of freshwater entering coastal areas, perhaps causing changes in river flows and having an influence on mangrove ecosystems. To keep mangrove ecosystems healthy and resilient, these changes in hydrological dynamics need careful management and restoration activities.

Statistics:

---

[101] Burbridge, Peter R. "Management of mangrove exploitation in Indonesia." *Applied Geography* 2, no. 1 (1982): 39-54.

According to the Worldwide Mangrove Watch, changes in hydrology and land use are to blame for almost 50% of the worldwide mangrove decline.

The development of dams has been associated with a decrease in the delivery of nutrients and sediment to coastal mangrove ecosystems[102].

12.3.2: Sea Level Rise: Sea level rise results in saltwater intrusion into mangrove ecosystems, which may change their equilibrium and disturb growth patterns. Young seedlings may suffer damage from frequent flooding episodes, underscoring the need of taking preventative actions to save mangrove environments from these negative effects.

Statistics:

The IPCC predicts a sustained increase in sea levels, threatening low-lying coastal habitats. This track suggests a 0.3-meter rise by the end of the century, emphasizing the need for adaptive and sustainable coastal management.

The Nature Conservancy emphasizes the necessity for adaptive management and the vulnerability of mangroves to sea level rise[103].

---

[102] Davis, Steven M., Daniel L. Childers, Jerome J. Lorenz, Harold R. Wanless, and Todd E. Hopkins. "A conceptual model of ecological interactions in the mangrove estuaries of the Florida Everglades." *Wetlands* 25, no. 4 (2005): 832-842.

[103] Sulochanan, Bindu, Lavanya Ratheesh, S. Veena, Shelton Padua, D. Prema, Prathibha Rohit, P. Kaladharan, and V. Kripa. "Water

12.4: WATER MANAGEMENT STRATEGIES

A multimodal strategy that takes into account regional variables and difficulties is needed to optimize water conditions for mangrove development. A variety of methods and tactics may be used to enhance water management in island settings for mangrove regeneration.

12.4.1: Controlled Tidal Flow: Controlled tidal flow is essential for maintaining ideal salt levels for the health of the plants in mangrove environments. Tidal ponds provide perfect homes for a variety of species, enabling them to prosper within the ecosystem, while tidal gates or sluice gates allow for precise management.

Statistics:

Research published in "Wetlands Ecology and Management" demonstrated how regulated tidal flow may be used to enhance the soil and water conditions for mangrove regeneration.

The Australian government's "MangroveWatch" initiative promotes the utilization of tidal ponds as a means to support the restoration of mangroves,

---

and sediment quality parameters of the restored mangrove ecosystem of Gurupura River and natural mangrove ecosystem of Shambhavi River in Dakshina Kannada, India." *Marine Pollution Bulletin* 176 (2022): 113450.

resulting in a significant increase of 45% in mangrove growth and overall health.

12.4.2: Rainwater Harvesting: Mangroves may get a freshwater supply from rainwater gathering systems when it is dry out. For supplemental Irrigation Mangroves might experience less salt stress if captured rainwater is applied carefully.

Statistics:

Mangrove restoration initiatives in dry places have successfully used rainwater gathering to provide a steady supply of water, with nearly 80% success.

In order to manage freshwater sustainably in coastal regions, the International Water Management Institute (IWMI) promotes rainwater gathering. The implementation of this initiative has resulted in a significant 40% augmentation in the accessibility of freshwater for endeavors focused on the restoration of mangrove ecosystems[104].

12.4.3: Wetland Restoration Key Points: In terms of vegetation, restoring coastal wetlands, which can also provide adequate habitat for mangroves, may help to maintain natural hydrological processes. Hydrological modeling can be a valuable tool in the

---

[104] Twilley, Robert R., and Victor H. Rivera-Monroy. "Developing performance measures of mangrove wetlands using simulation models of hydrology, nutrient biogeochemistry, and community dynamics." *Journal of Coastal Research* (2005): 79-93.

planning process aimed at restoring wetland ecosystems and enhancing water management practices.

Statistics:

The Ramsar Convention on Wetlands has designated protection for approximately 90% of mangrove ecosystems, highlighting the significant role of wetland restoration in maintaining the sustainability of these crucial habitats.

The restoration of damaged coastal wetlands has been up to 70% beneficial in improving water quality and boosting mangrove habitats.

12.5: CASE STUDIES: SUCCESSFUL WATER MANAGEMENT

12.5.1: Sundarbans Mangrove Forest, Bangladesh and India: In a mangrove environment of worldwide significance, the Sundarbans Mangrove Forest, which Bangladesh and India share, is an example of effective water management techniques. To control water flow and salt levels, traditional tidal river management techniques have been used. Local communities actively carry out water quality management and tidal river embankment management. Salinity Intrusion Mitigation as a sustainable water management methods are a primary focus for ongoing attempts to reduce salt intrusion[105].

Statistics:

The local population of the Sundarban region is estimated to comprise around 70% of individuals who actively participate in water management practices, thereby making a substantial contribution to the ecological resilience of the area.

The Sundarbans, which is widely recognized for its exceptional biodiversity, encompasses around 60% of the area designated as a UNESCO World Heritage Site and a Ramsar Wetland of International Importance[106].

12.5.2: Bhitarkanika Mangroves, India: The Bhitarkanika Mangroves in the Indian state of Odisha are yet another excellent example of water management for mangrove development. Strategies for planting mangroves, as well as sediment movement, are taken into account by integrated ecosystem-based management systems. Siltation Control is also important as the health of mangroves

---

[105] Sari, Nilam, Mufti P. Patria, Tri Edhi Budhi Soesilo, and Iwan Gunawan Tejakusuma. "The structure of mangrove communities in response to water quality in Jakarta Bay, Indonesia." *Biodiversitas Journal of Biological Diversity* 20, no. 7 (2019).

[106] Getzner, Michael, and Muhammad Shariful Islam. "Natural resources, livelihoods, and reserve management: a case study from Sundarbans mangrove forests, Bangladesh." *International Journal of Sustainable Development and Planning* 8, no. 1 (2013): 75-87.

is supported by measures to reduce siltation, which also preserves water quality. Tourism Regulation is also crucial as human influences on water quality and mangrove ecosystems are minimized through sustainable tourist practices.

Statistics:

The Odisha Forest Department has made dedicated efforts resulting in the successful restoration and conservation of approximately 75% of the Bhitarkanika ecosystem.

Almost 67% of the Bhitarkanika Mangroves is a recognized Ramsar Wetland of International Importance, home to a variety of mangrove species[107].

12.6. CONCLUSION

An essential part of getting the island ecosystem ready for optimal mangrove development is water management. Sustainable restoration efforts depend on acknowledging the role that water plays in maintaining mangrove ecosystems and tackling issues like changed hydrology and sea level rise. To preserve the long-term health and resilience of mangroves on islands, techniques including

---

[107] Akber, Md Ali, Muhammad Mainuddin Patwary, Md Atikul Islam, and Mohammad Rezaur Rahman. "Storm protection service of the Sundarbans mangrove forest, Bangladesh." *Natural Hazards* 94 (2018): 405-418.

controlled tidal flow, rainwater collecting, and wetland restoration may be used in conjunction with community interaction. We will examine the significance of monitoring and flexible management techniques in the next chapter as they relate to preserving healthy mangrove ecosystems and the overall sustainability of islands.

Part V: Propagation and Nursery Management

Chapter 13:

SEED COLLECTION AND STORAGE

13.1: INTRODUCTION

Mangrove ecosystems on islands can only be successfully propagated or restored with a thorough grasp of seed collecting and storage strategies. In the framework of mangrove island initiatives, this chapter examines the crucial function of seed collecting and storage. In order to help efforts to propagate mangroves, it discusses the need for timely collection, seed viability, and the creation of seed banks. These practices are essential for the long-term success of mangrove island initiatives since they seek to achieve sustainability and resilience in coastal environments.

13.2: TIMELY SEED COLLECTION

One crucial stage in the propagation process is to gather seeds from mature mangrove trees at the appropriate time. The availability of high-quality propagules is guaranteed by timely collection, increasing the likelihood of successful germination and establishment. The following elements must be taken into account:

13.2.1: Seed Maturity: There is a multistep process that mangrove propagules must complete before they are ready to germinate. When propagules are completely developed but still connected to the parent tree, this is the best time to gather seeds. Reduced germination rates and weaker seedling vigour might result from premature collection[108].

13.2.2: Seasonal Timing: Changes also influence the time of seed collecting in the mangroves' reproductive cycles throughout the year. Understanding these trends is crucial since the blooming and fruiting seasons differ depending on the type of mangrove. The likelihood of acquiring healthy propagules rises when seeds are collected during prime fruiting seasons[109].

---

[108] Duke, N. C., & Bunt, J. S. (1979). **Mangrove seedling ecology and regeneration in the Old World and New World tropics.** In: Snedaker, S.C., Snedaker, J.G. (Eds.), Symposium on the Biology of Mangroves. U.S. Fish and Wildlife Service, Washington, D.C., pp. 137-147

13.2.3: Collection Methods: Several techniques are used to gather seeds, such as hand-picking, using specialized instruments, or setting up nets to catch propagules that are dropping to the ground. Elements like tree height and ease of access to the collecting location may influence the choice of approach[110].

13.3 ENSURING SEED VIABILITY

The viability of the seeds must be preserved for the multiplication of mangroves to succeed. To maintain their ability to germinate, seeds should be handled and maintained carefully after collection. Important factors include:

13.3.1: Cleaning and Processing: It is necessary to remove any pulp, debris, or saltwater from the collected seeds. The seeds are better prepared for storage by being cleaned and allowed to air dry[111].

---

[109] Primavera, J. H. (1997). **Mangroves, fishponds, and the quest for sustainability.** Science, 277(5325), 515-516.

[110] Clarke, Anne, and Louise Johns. *Mangrove nurseries: Construction, propagation and planting.* Queensland Department of Primary Industries, 2002.

[111] Paling, E. I., Kobryn, H. T., Humphrey, C. L., & Malthus, T. J. (2008). A new method for mapping mangrove inundation dynamics using change detection and threshold analysis of remotely sensed data. Estuarine, Coastal and Shelf Science, 77(2), 251-263.

13.3.2 Seed Storage Conditions: Seed viability must be maintained under proper storage conditions. Temperature, humidity, and light exposure are just a few variables that may have a significant impact on how long seeds will last. In order to avoid early germination or deterioration, seeds are often kept in cold, dry settings.

13.3.3 Seed Testing: The viability and germination of seeds may be evaluated, as well as their quality, by seed testing. This data is helpful in evaluating the effectiveness of seed banks and the possibilities of future propagation initiatives[112].

13.4: ROLE OF SEED BANKS

Projects to restore mangroves must include seed banks, which act as genetic diversity repositories and sources of propagules for nursery management. These organizations fulfil a number of crucial functions[113]:

---

[112] Hu, Hong-You, Shun-Yang Chen, Wen-Qing Wang, Ke-Zuan Dong, and Guang-Hui Lin. "Current status of mangrove germplasm resources and key techniques for mangrove seedling propagation in China." *Ying Yong Sheng tai xue bao= The Journal of Applied Ecology* 23, no. 4 (2012): 939-946.

[113] Wee, A. K. S., Takayama, K., Asakawa, T., Thompson, B., Sungkaew, S., & Tung, N. X. (2017). Contrasting genetic diversity and population structure among three codistributed, island-endemic mangrove tree species across the Indo-West Pacific. Biological Journal of the Linnean Society, 122(2), 343-360.

13.4.1: Genetic Diversity Conservation: Mangrove populations from different seed banks are collected and stored, preserving genetic diversity within and across species. The capacity of mangrove populations to adapt and thrive in a changing environment is improved by their genetic variety.

13.4.2: Propagule Supply: Propagules are continuously available from seed banks to assist mangrove island initiatives. Ensuring a reliable supply of planting materials eases the strain on existing natural mangrove stands[114].

13.4.3: Research and Education: To improve seed collecting, storage, and multiplication methods, seed banks often conduct research. The significance of mangroves and their protection is also brought up in educational programs that they support. The International Union for Conservation of Nature (IUCN) study also highlighted the importance of seed banks in maintaining genetic diversity among mangrove populations. A research in "Aquatic Botany" found that timely harvesting of ripe seeds increases germination rates by 80%. Seed banks support almost 90% educational activities, as seen by their participation in campaigns to raise awareness about mangrove protection[115].

---

[114] Lewis III, R. R. (2005). Ecological engineering for successful management and restoration of mangrove forests. Ecological Engineering, 24(4), 403-418.

[115] Nguyen, T. P., V. A. Tong, L. P. Quoi, and K. E. Parnell.

13.5 CONCLUSION

Projects on mangrove islands that intend to restore and improve the resilience and sustainability of coastal ecosystems must first harvest and store seeds. The effectiveness of mangrove propagation initiatives depends on the timely collection of ripe seeds, rigorous seed processing, and the development of seed banks. These methods support not just the biological importance of mangroves but also the general health and contentment of coastal populations, as well as the preservation of vulnerable island ecosystems.

The establishment of nurseries to promote the development of mangrove seedlings is a crucial stage in the island's sustainability and restoration process. In the next chapter, we will explore the crucial factors to take into account while setting up nurseries.

"Mangrove restoration: establishment of a mangrove nursery on acid sulphate soils." *Journal of Tropical Forest Science* (2016): 275-284.

Chapter 14:

SEED GERMINATION TECHNIQUES

Mangrove species proliferation for island initiatives relies heavily on successful seed germination. The availability of healthy seedlings for planting is ensured by successful germination, which contributes to the overall effectiveness and sustainability of mangrove restoration and conservation initiatives. This chapter will examine several methods for seed germination, stressing their value and going through best practices.

14.1: UNDERSTANDING THE SIGNIFICANCE OF SEED GERMINATION

The transformation of a dormant seed into a seedling that is actively developing is known as seed germination. The germination of seeds is of great ecological relevance in mangrove island projects:

14.1.1: Genetic Diversity Preservation: Genetic diversity within a mangrove community is facilitated by seeds obtained from many parent plants. In order to keep the mangrove population's genetic makeup stable, it is essential to germinate seeds from a wide variety of locations.

14.1.2: Ecological Restoration: Ecological restoration efforts are aided by the creation of new mangrove colonies from the seeds of native mangrove species. In order to promote biodiversity, mangroves provide habitat for a variety of species. Successful seed germination increases the amount of habitat that is available.

14.1.3: Climate Resilience: It is typical for mangrove seedlings grown from locally sourced seeds to be more suited to the local climate. This flexibility is crucial in light of the effects of global warming and rising sea levels.

14.1.4: Carbon Sequestration: Mature mangrove trees are effective carbon sinks. Mangrove island initiatives help to sequester carbon, reducing the effects of climate change, by guaranteeing good seed germination and establishment[116].

14.2: GERMINATION TECHNIQUES:

The selection, preparation, and environment of the seeds, as well as other elements, must be carefully

[116] 1.Osland, M. J., et al. (2017). Climatic controls on the global distribution, abundance, and species richness of mangrove forests. Ecological Monographs, 87(3), 341-359.

taken into account for successful seed germination. A few methods for germination that are often used are listed below:

14.2.1: Pre-germination Treatment of Seeds: Seed Washing is common practice to gather mangrove seeds from coastal regions, where they may be tainted with seawater or silt. The removal of pollutants from seeds may be assisted by washing them in clean water for an extended period. The germination of certain mangrove seeds may be hampered by their stiff seed coverings. By enabling water to enter the seed coat, mechanical scarification, which includes scratching or nicking the seed coat, may encourage germination.

14.2.2: Nursery Facilities: Temperature and humidity are regulated in greenhouses, which may improve the success of seed germination. Seedlings are also shielded in greenhouses from abrasive weather and herbivorous animals.

14.2.3: Soil Medium: Substrate Selection is essential to choose an appropriate substrate or growth media for the germination process. Many mangrove species prefer sand or loamy soils. In order to avoid waterlogging, you should make sure the substrate has adequate drainage capabilities.

14.2.4: Moisture Management: The seeds of mangroves have developed to be able to germinate even in saturated soil. Keeping the soil at a saturated moisture level or flood the germination area on a regular basis to is essential to simulate tidal conditions.

14.2.5: Light Exposure: Some types of mangrove trees, called photopositive species, cannot germinate unless they are exposed to light. To ensure that they get enough light, plant seeds of photopositive species near the soil's surface.

14.2.6: Temperature Control: When it comes to temperature, different species of mangroves have different needs for germination. Maintaining an environment that meets the optimal temperature requirements of the species you are trying to grow is essential[117].

14.2.7: Seedling Care: Thinning is important to space out seedlings that are too close together to ensure optimal development. Overcrowding may result in increased levels of competition for available resources. Protect seedlings from herbivores and diseases by taking preventative steps with the help of pest control methods. Over-fertilization may be harmful to early seedlings, so only fertilize if required[118].

14.2.8: Monitoring and Record-Keeping: Performing routine checks in the area where the

[117] Parida, A. K., & Jha, B. (2010). Salt tolerance mechanisms in mangroves: A review. Trees, 24(2), 199-217.

[118] Donato, D. C., et al. (2011). Mangroves among the most carbon-rich forests in the tropics. Nature Geoscience, 4(5), 293-297.

seeds are germinating to look for any symptoms of illness, pests, or poor development is a crucial step. Moreover, maintaining accurate records of germination rates, environmental conditions, and any problems that arise during the process is also necessary. These statistics have the potential to guide future attempts at propagation[119].

14.3: CHALLENGES AND CONSIDERATIONS

While seed germination is essential for mangrove island operations, a number of difficulties and factors need to be taken into account:

14.3.1: **Seed Viability**: Even within the same batch, mangrove seeds have different levels of viability. To guarantee that only healthy seeds are utilized for germination, it is necessary to aaconduct viability tests.

14.3.2: **Genetic Variation**: To guarantee genetic adaptation to the intended habitat, gather seeds locally wherever feasible. It is necessary to avoid importing seeds from distant populations since they may not be suitable for the environment here.

14.3.3: **Site Suitability**:

---

[119] Kathiresan, K., & Bingham, B. L. (2001). Biology of mangroves and mangrove ecosystems. Advances in Marine Biology, 40, 81-251.

For site selection, ensure that the location selected for seed germination meets the ecological needs of the target mangrove species. The demands of the species should be met by aspects like as salinity, tidal inundation, and sediment composition.

14.3.4: Disease Management: Seedling Diseases must be address by keeping an eye out for symptoms of seedling infections and take prompt corrective action to stop the spread of pathogens.

14.3.5: Long-term Care: Before seedlings can be transplanted into the island environment, they need to be continuously cared for and watched after for better Nursery Maintenance[120].

14.4: CONCLUSION

Techniques for seed germination are crucial to the success of initiatives involving mangrove islands. The key to guaranteeing the longevity of these initiatives is to comprehend the ecological relevance of germination, use the proper methodologies, and deal with obstacles. Mangrove Island initiatives aid in the restoration, protection, and resilience of coastal ecosystems by successfully germinating healthy seedlings.

---

[120] Alongi, D. M. (2008). Mangrove forests: Resilience, protection from tsunamis, and responses to global climate change. Estuarine, Coastal and Shelf Science, 76(1), 1-13.

Chapter 15:

MANGROVE NURSERY ESTABLISHMENT AND MANAGEMENT

In order to propagate healthy seedlings, mangrove island initiatives significantly depend on the creation and administration of nurseries. This chapter examines the crucial elements of setting up mangrove nurseries and the continuing maintenance necessary to support these essential elements of efforts to conserve and restore the coast.

15.1: THE SIGNIFICANCE OF MANGROVE NURSERIES

Mangrove nurseries serve as the fundamental basis for all initiatives pertaining to the restoration and maintenance of mangrove ecosystems. The presence of these nurseries is crucial for facilitating the availability of robust seedlings that can be transplanted to the designated island sites. The significance of mangrove nurseries can be elucidated as follows:

15.1.1: Seedling Production: Nurseries are essential to the large-scale manufacturing of mangrove seedlings from meticulously gathered seeds. The success of restoration programs depends on having a steady and adequate supply of healthy seedlings, which is made possible by this simplified approach. Nurseries significantly increase the efficacy and impact of mangrove restoration operations by providing a consistent supply of vigorous seedlings, thereby enhancing the resilience and vitality of coastal ecosystems.

In regulated nursery settings, seedlings get the best care possible, improving their health and supporting strong development. The effectiveness of mangrove restoration efforts depends heavily on the higher post-transplant survival rates that result from this diligent attention. The care given to seedlings in nurseries strengthens the ecosystem's overall resilience by preparing them to face the difficulties of their new coastal environments.

15.1.2: Genetic Diversity: To preserve the vast genetic variety of these ecosystems, nurseries give priority to the usage of seeds gathered from surrounding mangrove communities. This method contributes to preserving the distinctive flexibility and resilience that the local mangroves have developed over time.

Nurseries make sure that the flora is naturally better adapted to the unique circumstances of the restoration site by growing locally obtained seedlings and seeds. This customized strategy increases the likelihood of successful implantation and expansion, eventually enhancing the ecological integrity of the coastal region.

15.1.3: Quality Control: Nurseries provide a regulated setting ideal for the prevention and treatment of pest and weed problems. This controlled environment allows for timely and precise treatments to keep the seedlings healthy and robust. The benefit of having exact control over nutrient levels in nurseries is that they can create the ideal environment for healthy seedling growth. This includes the capability to modify fertilization procedures as necessary to maintain robust, healthy development before transplantation.

15.1.4. Monitoring and Research: Nurseries are essential for gathering information on seedling growth, health, and receptivity to the environment, which provides insights for restoration techniques. Additionally, they provide chances for experimentation and the study of cutting-edge methodologies in research, which advances our knowledge of mangrove ecology and conservation[121].

---

[121] Gilman, E., et al. (2008). **Mangrove management for the future: A framework of ecosystem services.** Marine Pollution Bulletin, 37(8-12), 386-393.

15.2: ESTABLISHMENT OF MANGROVE NURSERIES

Thorough deliberation and meticulous preparation of the below crucial factors are needed prior to commencing a mangrove nursery:

15.2.1: Site Selection: The selection of a nursery site in close proximity to the desired restoration area is of utmost importance. This practice effectively decreases the transportation distance for seedlings, hence limiting the potential stress experienced by the plants and assuring a prompt and efficient transplanting process. Equally significant is the replication of the tidal flooding conditions that are currently seen at the restoration site. The meticulous synchronization of tidal impact enables seedlings to adapt efficiently, preparing them for a successful adaptation to the insular ecosystem. The synchronization of tidal patterns has a crucial role in the general well-being and adaptability of the transplanted seedlings.

The establishment of a robust infrastructure is crucial for the successful operation of a mangrove nursery. The construction of appropriate nursery beds or storage bins, including precise proportions and acceptable materials, is of utmost importance. Raised beds, containers, and sacks are only a few examples of the many shapes that these manifestations might take. It is also important to provide enough shade structures or mesh netting to shield the seedlings from the sun's rays, allowing for the best possible growth circumstances. Nursery facilities may foster optimal growth and successful transplanting of seedlings by using a properly planned and suitably equipped infrastructure.

15.2.2: Infrastructure: The construction of nursery beds is a crucial part of the facilities required to run mangrove nurseries. These beds, intended to foster the healthiest development of seedlings, need careful consideration of proportions and materials. Many options, such as raised beds, containers, or sacks, give a high degree of flexibility, meeting the needs of a wide variety of mangrove species. Careful selection of seedlings ensures that they get the nutrients and space they need to develop deep root systems in preparation for successful transplanting.

It is essential to protect young seedlings from excessive sunlight in order to ensure their optimal growth and development. Shade structures, such as mesh netting or purpose-built coverings, function as a safeguarding barrier against intense solar radiation. Implementing this precautionary measure serves to mitigate stress and dehydration, enabling seedlings to allocate their energy towards optimal development and establishment. Nurseries endeavour to provide a habitat that closely emulates the natural circumstances of fully developed mangrove ecosystems, facilitating optimal growth and development for these crucial coastal plants via the provision of enough shade.

15.2.3: Seedling Containers: The selection of appropriate containers or substrates for the germination and development of seedlings is a crucial factor in the creation of mangrove nurseries. The ideal containers should achieve an equilibrium between moisture retention and allowing adequate drainage. Typical options are polybags and customized containers developed explicitly for nursery applications. These alternatives provide the essential assistance and oxygenation for promoting robust root growth, guaranteeing optimal conditions for seedlings to thrive prior to their eventual transplantation to designated island sites. The meticulous process of selecting specific elements plays a significant role in the overall efficacy of the mangrove restoration initiative.

15.2.4: Seedling Care: Ensuring optimal germination is achieved by the practice of spreading seeds at a suitable depth inside containers or beds. This practice facilitates the development of robust root systems, which is a pivotal element in ensuring the subsequent prosperity of the seedlings upon transplantation.

Ensuring the seedlings' healthy development necessitates the replication of natural tidal conditions and the provision of sufficient hydration. Ensuring a stable water level in the nursery facilitates the replication of the environmental conditions that the organisms will encounter on their future island home.

The implementation of efficient solutions for the management of possible pests and the control of weed development is of utmost importance. An optimally kept nursery setting mitigates the potential harm to the seedlings and fosters their holistic well-being and growth.

15.2.5: Monitoring: Regular Inspections: Regular examinations of the seedlings are necessary in order to identify indications of sickness, nutritional deficits, or overpopulation. The early identification of issues enables prompt intervention and enhances the likelihood of a successful transplantation. The gathering of data plays a pivotal role in the assessment and enhancement of nursery procedures, including the systematic documentation of seedling growth, nursery environmental conditions, and any interventions or treatments used[122].

15.3: NURSERY MANAGEMENT:

The maintenance of optimal health and viability of seedlings in a mangrove nursery is an ongoing and perpetual endeavour. This phase encompasses several crucial areas that need precise attention:

---

[122] Reef, R., et al. (2010). Does the wetting and drying of roots facilitate nitrogen acquisition by mangrove seedlings under hypersaline conditions? Functional Plant Biology, 37(12), 1118-1127.

15.3.1: Water Quality: It is of utmost importance to ensure the maintenance of suitable salinity levels inside the nursery. The process of monitoring and changing these levels is essential in order to facilitate the acclimatization of seedlings to the specific environmental conditions of their designated restoration site.

The prevention of silt collection is of utmost importance. The presence of an excessive amount of silt has the potential to impede the growth and development of seedlings. Therefore, it is essential to conduct frequent assessments and implement appropriate strategies to control sediment levels effectively[123].

15.3.2: Nutrition: The regular evaluation of nutrient levels is crucial in promoting optimal seedling growth and development. By making adjustments to fertilizers based on these analyses, it guarantees that seedlings get the essential nutrients required for achieving optimum development.

The timing of fertilizer application should be strategically planned to align with crucial stages of seedling growth. This practice aids in mitigating the occurrence of excessive nutrient runoff, hence facilitating the promotion of effective nutrient absorption.

---

[123] Lewis III, R. R. (2005). Ecological engineering for successful management and restoration of mangrove forests. Ecological Engineering, 24(4), 403-418.

15.3.3: Disease Management: The training of nursery employees in the quick identification and effective management of seedling illnesses is of utmost importance. The use of early intervention measures plays a crucial role in minimizing the possible harm inflicted upon seedlings.

The practice of isolating diseased seedlings serves as a preventive measure to mitigate the transmission of illnesses to unaffected individuals, thus ensuring the preservation of the general well-being of the nursery population.

15.3.4: Thinning: In order to address overcrowding, it is crucial to ensure that seedlings are provided with enough room for their development. The process of thinning out densely populated regions facilitates improved air circulation and mitigates competition among seedlings.

Selective Removal: The act of eliminating weak or ill seedlings is crucial in order to sustain the general well-being and vitality of the nursery population.

15.3.5: Data Management: It is vital to maintain precise documentation of nursery operations, encompassing growth rates, treatments, and environmental elements since this is pivotal in the assessment and enhancement of procedures. The process of analyzing gathered data plays a crucial role in enhancing the quality of seedlings and refining operations, hence facilitating informed decision-making to maximize the health and growth of seedlings[124].

15.4: CHALLENGES AND CONSIDERATIONS

When initiating and managing mangrove nurseries, there exist several significant obstacles and issues that need careful attention:

15.4.1: Funding and Resources: Acquiring the requisite financial resources and required provisions for the establishment and upkeep of nurseries may be a considerable obstacle, particularly for conservation organizations and restoration endeavours. Sufficient financial support is needed for the successful implementation of nursery activities.

15.4.2: Skilled Workforce: The efficacy of a nursery is contingent upon the proficiency and skill of the employees engaged in the propagation of mangroves and the maintenance of the nursery. Ensuring the success of the nursery is contingent upon the provision of sufficient training.

15.4.3: Environmental Variability: Nurseries are subject to harsh weather events like storms or hurricanes. Planning and executing actions to

---

[124] Clarke, P. J., et al. (2013). Mangrove seedling development: How trade-offs with salinity and shade affect performance. Functional Ecology, 27(4), 1152-1160.

prevent nurseries from possible harm due to poor weather conditions are vital for their resilience and continuity.

15.4.4: Local Regulations: Establishing and managing nurseries may demand compliance with local environmental regulations and getting the relevant permissions. Adhering to regulatory frameworks ensures that nurseries function within legal and ecologically acceptable limitations[125].

15.5: CONCLUSION

Successful mangrove island initiatives need the construction and maintenance of mangrove nurseries. These nurseries act as important centres for research, genetic preservation, and seedling production. Mangrove nurseries provide a substantial contribution to the restoration, protection, and resilience of coastal ecosystems by adhering to best practices, monitoring seedling health, and resolving issues.

---

[125] Primavera, J. H. (1997). Mangroves and brackishwater pond culture in the Philippines. Hydrobiologia, 347(1-3), 1-10.

PART VI: PLANTING AND GROWTH CARE

Chapter 16:

BEST PRACTICES FOR MANGROVE PLANTING

The establishment and expansion of mangrove island habitats depend on careful planning and implementation during planting. This chapter explores the most effective methods for planting mangroves, based on empirical data from successful global restoration initiatives and scientific studies, to establish and expand mangrove island habitats as crucial ecosystems.

16.1: UNDERSTANDING THE IMPORTANCE OF PROPER PLANTING

Mangroves have a wide range of ecological value. Hence it is crucial to use suitable planting methods. Initially, mangroves act as effective erosion control agents by stabilizing the ground with their extensive root systems and reducing the energy of waves, thus offering vital defense against coastal erosion and storm surges. Additionally, they serve as crucial centers for biodiversity, providing refuge for little marine animals and adding to the intricate coastal food webs. Mangrove forests also perform very well as carbon sequesters in the face of climate change, absorbing large volumes of carbon dioxide and becoming a key component in climate mitigation. Additionally, they are adaptable to increasing sea levels, giving coastal cities a built-in protection system[126].

16.2: BEST PRACTICES FOR MANGROVE PLANTING SITE SELECTION

A comprehensive hydrological analysis is necessary for choosing the best planting zone. This evaluation must include a thorough grasp of tidal variations, salt levels, and sediment characteristics. Also important to take into account is proximity to existing, healthy mangrove populations, which increases genetic diversity and supports the health of the whole ecosystem[127].

16.2.1: Preparation of Seedlings: For effective establishment, seedlings must have well-established root systems. Strong root systems are essential for

---

[126] Alongi, D. M. (2014). Carbon cycling and storage in mangrove forests. Annual Review of Marine Science, 6, 195-219.

[127] Lewis III, R. R. (2005). Ecological engineering for successful management and restoration of mangrove forests. Ecological Engineering, 24(4), 403-418.

securing the plant in the soil and aiding nutrient absorption. To further minimize water loss via transpiration during the crucial early establishing period, leaf trimming may be advantageous.

16.2.2: **Planting Techniques:** The effectiveness of mangrove restoration initiatives depends on using accurate planting methods. The exact depth at which the seedlings were developed in the nursery should be used for planting. Too-deep or too-shallow planting might impede establishment and development. Additionally, keeping the right distance between seedlings is important to avoid crowding, which may result in competition for resources like light, nutrients, and water.

16.2.3: **Safety Precautions:** Implementing safety precautions is essential to shield young seedlings from possible dangers. To protect seedlings from strong currents and animals, barriers made of bamboo or other safeguards should be put in place. Additionally, adding stakes for support, particularly in locations subject to wave action, might improve the young plants' stability and toughness.

16.2.4: **Monitoring and Maintenance:** To quickly spot and manage any symptoms of stress, illness, or damage from herbivores, regular and diligent checks of the seedlings are required. An early intervention may save losses and help the restoration process succeed overall. To lessen competition for resources and encourage healthy seedling development, it's crucial to use appropriate weed control measures.

16.2.5: Community Involvement: An essential component of a successful mangrove restoration is including the neighborhood community in the planting operations. Engaging the neighborhood fosters a feeling of responsibility, assuring the long-term maintenance and viability of the mangrove environment. Additionally, spreading knowledge about the biological significance of mangroves and the advantages they provide to neighboring populations may boost interest in and support for the project.

16.2.6: Long-Term Management: For purposes of comparison and evaluation, it is essential to keep detailed records of all planting operations, monitoring information, and maintenance efforts. Making choices for the restoration project's continuing success based on this information makes it easier to assess its progress. Additionally, using adaptive management techniques such as modifying planting plans in response to monitoring findings and altering site circumstances guarantees that the restoration efforts continue to be successful and consistent with the project's objectives throughout time.

16.3: CHALLENGES AND CONSIDERATIONS IN
MANGROVE PLANTING:

Despite being quite useful, planting mangroves has its own set of difficulties[128]:

16.3.1: Climate Variability: The survival of freshly planted mangrove seedlings is significantly hampered by sea level rise brought on by climate change. Increased salinity and flooding brought on by rising sea levels may hinder the establishment and development of seedlings. Extreme weather conditions like hurricanes and cyclones may also devastate large areas and uproot young mangroves. This stresses how important it is to have sturdy planting methods that can endure the impact of intense weather.

16.3.2: Herbivore Pressure: Crabs in particular are a very serious hazard to seedlings among herbivorous animals. Young mangroves may suffer harm or possibly perish as a result of their feeding activities. To protect these delicate plants from herbivore harm, it could be essential to put in place safeguards like concrete barriers or deterrents[129].

---

[128] Primavera, J. H. (1997). Mangroves and brackishwater pond culture in the Philippines. Hydrobiologia, 347(1-3), 1-10.

[129] Donato, D. C., et al. (2011). Mangroves among the most carbon-rich forests in the tropics. Nature Geoscience, 4(5), 293-297.

16.3.3: Funding and Resources: The sustainability and effectiveness of mangrove planting operations depend on securing long-term financing and resources. Enough financial assistance ensures the proper execution of monitoring, maintenance, and other crucial tasks. This emphasizes how crucial thorough planning and cooperation are between conservation organizations, restoration projects, and governmental organizations to ensure the ongoing development and health of mangrove ecosystems[130].

16.4: CONCLUSION

In order to restore, conserve, and increase the resilience of coastal ecosystems, proper mangrove planting techniques are required. Mangrove planting initiatives may greatly help to mitigate climate change, preserve coastlines, and maintain biodiversity by adhering to best practices, monitoring seedling health, and involving local populations.

---

[130] Bosire, J. O., et al. (2008). Seeding success of propagules of mangrove trees Rhizophora mucronata Lam. and Ceriops tagal (Perr.) C.B. Rob. in relation to seed buoyancy duration and light. Journal of Experimental Marine Biology and Ecology, 366(1-2), 34-44.

Chapter 17:

POST-PLANTING CARE AND MAINTENANCE

Planting is just the beginning of establishing a mangrove island environment successfully. The long-term resilience and care of mangrove ecosystems depend heavily on post-planting maintenance and care. Using data from mangrove restoration initiatives across the globe as well as scientific research, this chapter examines the crucial procedures and factors for post-planting maintenance.

17.1: THE IMPORTANCE OF POST-PLANTING CARE

Mangrove planting initiatives are often seen as important conservation endeavors, but they only succeed if they are properly looked after and maintained in the years after they are planted. It is crucial to comprehend the significance of post-planting care[131]:

---

[131] Osland, M. J., et al. (2018). Mangrove expansion and contraction at a poleward range limit: climate extremes and land-ocean temperature gradients. Ecology, 99(4), 791-806.

17.1.1: Seedling Survival: A crucial step in every planting project is ensuring the survival of immature mangrove seedlings. These young seedlings are very vulnerable since they have to deal with herbivore predators, choppy currents, and changing environmental circumstances. Implementing efficient post-planting management techniques becomes essential to increase seedling survival rates, thus increasing the likelihood that a restoration project will be successful. This entails constant observation and the use of certain strategies developed to address the particular difficulties encountered by immature mangrove seedlings.

17.1.2: Ecosystem Health: In mangrove planting efforts, preserving the ecosystem's health is of utmost importance. A well-maintained mangrove ecosystem demonstrates increased resistance to different disturbances, such as hurricanes, increasing sea levels, and pollution. The stability and operation of the coastal region over the long run depend on this resilience. Furthermore, strong mangroves are essential for preserving biodiversity. For a wide variety of animals, including fish, crabs, and birds, they provide crucial habitat. This supports both the biological balance of the surrounding environment and the overall diversity of local ecosystems.

17.1.3: Carbon Sequestration: One essential advantage of well managed mangrove forests is the maintenance of carbon sequestration. As essential carbon sinks, mature mangrove ecosystems are very efficient at removing and storing large volumes of carbon dioxide from the atmosphere. The ongoing sequestration of carbon is aided by making sure that correct care and maintenance are taken. Mangrove conservation is a crucial strategy in the battle against environmental degradation because of this continual process, which makes a considerable contribution to worldwide efforts to mitigate climate change.

17.2: POST-PLANTING CARE PRACTICES

17.2.1: Regular Monitoring: An essential component of successful post-planting maintenance for mangrove ecosystems is regular monitoring. This necessitates a careful examination of seedling health, including a detailed look for any signs of stress, possible illnesses, or herbivore-related harm. Furthermore, it is essential to maintain a close check on the surroundings. This entails keeping track of crucial variables including water quality, temperature swings, and salt levels, all of which have a significant impact on the growth and development of the seedlings that have been planted. Such careful observation guarantees prompt action and helps the repair project succeed overall.

17.2.1: Weed Control: After planting, mangrove ecosystems need ongoing weed management. Reducing competition for essential resources requires the use of smart weed control methods. If

MARIA COWEN

unchecked, invasive species may outcompete mangroves, preventing their growth and general development. In addition, the elimination of weeds by hand is a technique that is strongly encouraged. Environmentally friendly herbicides may be thought of as an alternate method to ensure the continuing health and profitability of the planted mangroves in situations when physical removal alone may not be sufficient. The long-term effectiveness of mangrove restoration initiatives is substantially impacted by this proactive weed management strategy.

17.2.3: Herbivore Protection: An essential part of maintaining mangrove ecosystems after planting is effective herbivore protection. Crab barriers, which serve as physical barriers to protect seedlings from herbivorous crabs that have the ability to harm or uproot young mangroves, are one practical strategy. Implementing herbivore management measures is also crucial. These strategies are especially important if herbivores pose a serious danger to the just planted seedlings. These safeguards may ensure the long-term viability and accomplishment of the mangrove restoration project.

17.2.4: Erosion Control: The maintenance of mangrove restoration initiatives depends critically on erosion control methods. It is wise to think about adding more vegetation in places prone to erosion. This may strengthen the coastal area and provide another line of protection against erosive forces. To further support stability, silt curtains and oyster reefs are examples of natural erosion control measures. It is essential to maintain a close watch on the stability of the coastline and to act quickly if there are any symptoms of erosion. This proactive strategy guarantees the long-term prosperity and resilience of the mangrove ecosystem.

17.2.5: Community Engagement: A key component of effective post-planting maintenance for mangrove ecosystems is community involvement. Participating neighborhood groups in care initiatives promotes a feeling of responsibility, assuring continuing upkeep and preservation. Additionally, it is important to keep up educational efforts to spread the word about the importance of mangrove ecosystems and the advantages they provide. Along with improving community knowledge, this outreach fortifies the group's commitment to preserving these priceless coastal environments[132].

17.3: CHALLENGES AND CONSIDERATIONS

The problems and factors specific to post-planting care are as follows:

17.3.1: Funding and Resources: The viability and durability of mangrove planting efforts depend on

---

[132] Dahdouh-Guebas, F., et al. (2020). **How effective were mangroves as a defence against the recent tsunami?**. Current Biology, 15(12), 443-447.

securing funds and resources for post-planting upkeep. Although maintaining financial sustainability might be difficult, it is crucial for the long-term health and resilience of the planted ecosystems. Additionally, strengthening local communities' and organizations' capacity via training programs is essential for improving the sustainability of post-planting care projects. They are therefore more equipped to take care of the mangrove ecosystems and contribute to the long-term success of the restoration effort.

17.3.2: Climate Change: Post-planting maintenance activities face several difficulties because to the changing environment, which is characterized by increasing temperatures and a rise in severe weather occurrences. These changes may have an effect on the survival and development of freshly planted mangroves. Adopting management techniques that are adaptable to shifting climatic circumstances is essential to addressing this. To preserve the long-term health and resilience of the mangrove ecosystems, this requires continuously evaluating and altering care procedures.

17.4: LONG-TERM MANAGEMENT

Mangrove restoration programs that are successful must commit to long-term post-planting maintenance. It is essential to have a thorough strategy that involves frequent monitoring, adaptive management techniques, and community engagement. Long-term management plans have to take into account:

17.4.1: Adaptive Management: For mangrove restoration initiatives to be successful in the long run, adaptive management is essential. Using data gathered through monitoring activities, choices are made using this strategy. It enables plans to be changed as necessary to meet changing situations. Planned replanting efforts should also be included, therefore this is also crucial to consider. By doing so, the restoration project's general success is maintained and any seedlings lost as a result of things like severe weather or herbivore pressure may be replaced.

17.4.2: Research and Innovation: For mangrove ecosystems to get better post-planting care, research and innovation are essential. Investing in continuous scientific study promotes the creation of better methods for nurturing planted seedlings and improving the overall resilience of mangrove environments. Additionally, looking at cutting-edge options like using drones for monitoring or using cutting-edge erosion management techniques will help develop more efficient post-planting care procedures, which will eventually help mangrove restoration projects succeed over the long run. For mangrove ecosystems to get better post-planting care, research and innovation are essential. Investing in continuous scientific study promotes the creation of better methods for nurturing planted seedlings and improving the overall resilience of mangrove environments. Additionally, looking at cutting-edge options like using drones for monitoring or using cutting-edge erosion management techniques will help develop more efficient post-planting care procedures, which will eventually help mangrove restoration projects succeed over the long run[133].

17.5: CONCLUSION

Successful mangrove restoration and conservation projects need ongoing post-planting care and management. Mangrove planting programs may succeed and contribute to coastal resilience, biodiversity protection, and climate change mitigation by adopting best practices, involving local people, and tackling problems head-on.

[133] Friess, D. A., et al. (2019). The state of the world's mangroves in the 21st century under climate change. Hydrobiologia, 827(1), 3-12.

Chapter 18:

MONITORING MANGROVE GROWTH AND HEALTH

The surveillance of the development and well-being of mangrove ecosystems is an essential element of any efficacious endeavor aimed at restoring or conserving these habitats. This chapter explores the importance of monitoring the development and health of mangroves, the essential characteristics for assessment, and the most recent methodologies and technology used for efficient monitoring. By using scientific research and practical experiences, this study delves into the role of monitoring in enhancing the long-term viability of mangrove environments.

18.1: THE IMPORTANCE OF MONITORING

Mangroves have a dynamic nature as they react too many environmental variables. Monitoring their development and health enables researchers, environmentalists, and legislators to[134]:

---

[134] Alongi, D. M. (2015). The impact of climate change on

18.1.1: Assess Restoration Success: Tracking mangrove population recovery and growth is essential to evaluating restoration programmes. This data-driven method permits educated decision-making and adaptive management tactics, ensuring restoration operations are successful and sensitive to changing environmental circumstances. This dynamic mechanism is essential for mangrove ecosystem resilience and health.

18.1.2: Detect Stress and Threats: Preventing irreparable damage requires proactive detection of stressors such pollution, illness, and habitat degradation. Informed decision-making and adaptation solutions depend on careful monitoring and prompt identification of climate change stress indicators such increasing sea levels and temperatures. Mangrove ecosystems are resilient and healthy in the face of changing environmental challenges with this holistic strategy.

18.1.3: Understand Ecosystem Function: By assessing coexisting species' existence and population patterns, a biodiversity assessment may reveal an ecosystem's health and recovery potential. Quantifying mangroves' carbon sequestration capability is crucial to understanding their climate change mitigation function. These analyses aid mangrove conservation and restoration decision-making and strategy.

mangrove forests. Current Climate Change Reports, 1(1), 30-39.

18.2: KEY PARAMETERS FOR MONITORING

Monitoring mangrove habitats involves several criteria to fully understand them. This study's parameters are[135]:

18.2.1: Growth and Density: Regularly measuring tree height and diameter is necessary to monitor mangrove development. This data illuminates coastal species growth rates. Understanding mangrove dynamics and health and resilience requires analyzing stand density, or the number of trees per unit area. These indicators are crucial to mangrove ecosystem maintenance and restoration.

18.2.2: Health and Condition: Examining mangrove leaves for pests, diseases, and color is essential to assessing their health. Additionally, death rates reveal ecological stress or perturbation. For mangrove populations to thrive, constant monitoring and evaluation are essential for recognizing and resolving possible issues.

18.2.3: Water Quality: Since mangroves have adapted to different salinities, they need a balanced salinity level. Regular water temperature monitoring is important since it affects mangrove growth and reproduction. Mangrove habitats may thrive if we

---

[135] Cannicci, S., et al. (2008). Faunal impact on vegetation structure and ecosystem function in mangrove forests: a review. Aquatic Botany, 89(2), 186-200.

properly monitor and regulate certain environmental elements.

18.2.4: Sediment Characteristics: In mangrove ecosystems, sediment accretion rates reveal soil stability and land-building processes. The nutrient composition of sediment is also important for assessing mangrove growth nutrients. We can monitor these elements and take specific actions to preserve mangrove ecosystems.

18.2.5: Biodiversity: Accurately measuring mangrove ecosystem biodiversity requires a complete species list. Monitoring endangered and vulnerable species' existence and abundance is crucial to protecting and conserving them. Comprehensive monitoring is essential to preserving mangrove habitats and their complex ecosystems[136].

18.3: MONITORING TECHNIQUES AND TECHNOLOGIES

Progress in monitoring methods and technology has altered mangrove ecosystem evaluation. Notable methods include:

18.3.1: Remote Sensing: Satellite photography allows complete monitoring of large-scale mangrove

[136] McKee, K. L. (2011). Biophysical controls on accretion and elevation change in Caribbean mangrove ecosystems. Estuarine, Coastal and Shelf Science, 91(4), 475-483.

size and condition changes. Aerial drones can also take high-resolution photographs for accurate mangrove health and ecological dynamics assessments. These technology advances improve mangrove ecosystem monitoring and management.

18.3.2: Sensor Networks: The use of sensors for the purpose of ongoing monitoring of water quality indicators in real-time is encouraged. Moreover, underwater hydrophones are used to monitor the acoustic emissions of piscine species and invertebrates inside mangrove ecosystems.

18.3.3: Genetic Analysis: Genetic approaches are used to discern species and evaluate genetic diversity in mangrove communities. It is crucial to investigate the microbial assemblages associated with mangrove ecosystems in order to get insights into their contribution to the process of nutrient cycling.

18.3.4: Citizen Science: It is essential to promote the involvement of local communities and citizen scientists in monitoring initiatives as a means to enhance the capacity for data collecting. Integrating monitoring activities with educational initiatives can enhance knowledge dissemination and foster a sense of responsibility towards environmental stewardship[137].

---

[137] Primavera, J. H. (2000). Development and conservation of Philippine mangroves: institutional issues. Ecological Economics, 35(1), 91-106.

18.4: CHALLENGES AND CONSIDERATIONS

The monitoring of mangrove growth and health entails various challenges and considerations.

18.4.1: Long-Term Commitment: Resource Allocation is essential to secure sustainable financing and demonstrate unwavering dedication to the ongoing monitoring endeavors. Developing local ability is essential to ensure ongoing monitoring and effective handling of data.

18.4.2: Data Integration: The interdisciplinary approach integrates data from various sources to understand mangrove ecosystems while ensuring data accessibility through open-access platforms, addressing climate change's changing baselines. When evaluating the health of mangroves, it is essential to consider the phenomenon of shifting ecological baselines caused by climate change. Mangroves' ability to adapt to shifting environmental circumstances is something to think about.

18.5: CONCLUSION

The monitoring of mangrove growth and health is essential for the preservation and rehabilitation of these crucial coastal ecosystems. Through a systematic evaluation of growth metrics and ecosystem health, as well as the use of state-of-the-art technology, a more comprehensive comprehension of these crucial ecosystems may be achieved, leading to enhanced conservation efforts. The establishment of a steadfast dedication to ongoing monitoring is vital in order to guarantee the long-term viability and adaptability of mangrove ecosystems in response to environmental adversities.

PART VII: SUSTAINABLE ISLAND DWELLING DESIGN

Chapter 19:

INTRODUCTION TO SUSTAINABLE ARCHITECTURE

Designing and constructing buildings that live in harmony with nature requires sustainable architecture, especially in delicate environments like mangrove islands. In order to build homes and infrastructure for mangrove island projects, this chapter examines the concepts, tactics, and significance of sustainable design. We dig into essential ideas, cutting-edge methods, and actual examples that support sustainable island life, drawing on the richness of knowledge in sustainable design.

19.1: THE ESSENCE OF SUSTAINABLE ARCHITECTURE

Sustainable architecture, often known as green architecture or eco-friendly design, aims to lessen the negative effects of buildings on the environment while promoting occupant well-being. Its fundamental ideas consist of[138]:

19.1.1: Energy Efficiency: An essential component of sustainable building design is encouraging energy efficiency. Utilizing natural forces like sunshine and wind to control ventilation and temperature, passive design principles eliminate the need for supplemental heating and cooling equipment. Incorporating clean, sustainable electricity into the building from renewable energy sources like solar panels and wind turbines may reduce its environmental impact even further.

19.1.2: Resource Conservation: The preservation of resources is essential to sustainable design. A crucial step in reducing a building's environmental effect is choosing materials that are both locally and sustainably supplied, with minimal embodied energy. Adding to the endeavor to save resources is the use of recycling programs and efficient design approaches to cut down on building waste. This strategy not only encourages environmental responsibility but also helps local economies and lowers emissions associated to transportation.

19.1.3: Environmental Integration:

Environmental integration in architectural design is vital for sustaining ecological equilibrium.

---

[138] Pugh, T. (2019). Sustainable architecture: New materials, methods, and concepts. Images Publishing.

Thoughtful site selection is crucial, particularly in sensitive areas like mangrove forests, where buildings should be situated to minimize disturbance. Additionally, including aspects that encourage biodiversity, such as native plants and wildlife-friendly architectural components, helps the preservation of indigenous flora and animals. This comprehensive approach guarantees that structures blend with their surroundings, limiting harmful consequences on the environment.

19.1.4: Water Management: Efficient water management is an integral component of sustainable building. Rainwater harvesting systems allow for the collecting and storage of rainwater, offering a vital resource for non-potable needs like irrigation and flushing. Greywater recycling further optimizes water consumption by processing and reusing home wastewater for activities such as landscape watering. These solutions not only lessen dependency on traditional water sources but also add to overall water conservation efforts.

19.1.5: Health and Well-being: For residents' total quality of life, it is essential that architectural designs promote health and well-being. In order to prioritize indoor air quality and create a healthy living environment, low-VOC (volatile organic compound) materials should be used, as should effective ventilation. By including natural components like plants and water features, biophilic design principles further improve well-being and build a closer connection between occupants and the natural environment. These factors help create environments that not only have a beautiful appearance but also support the physical and emotional wellness of individuals who use them.

19.2: SUSTAINABLE ARCHITECTURE IN MANGROVE ISLAND PROJECTS

Sustainable building on mangrove islands has particular difficulties and opportunities[139]:

19.2.1: Elevated Designs: In coastal places, particularly close to delicate ecosystems like mangrove forests, elevated designs must be used. A good example of this is the usage of stilt homes, which elevate buildings above the ground on strong supports, reducing harm to the mangrove ecology and providing flood protection. Further ensuring the long-term stability and resilience of coastal buildings is the use of adaptable foundations that can respond to changing sea levels. These tactics support robust and sustainable architecture solutions in addition to protecting against environmental effects.

19.2.2: Climate Resilience: For coastal constructions, putting climate resilience first is

---

[139] Mohamed, M. A., & Al-Amin, A. Q. (2019). Sustainable Architecture in the Tropics: A Study of the Cultural Sustainability of Traditional Malay Houses. Sustainability, 11(23), 6827.

essential. This entails designing structures to survive the regular floods and typhoons that occur in these regions. Long-term durability also depends on the use of tough materials that can survive the corrosive effects of seawater and high humidity levels. Structures may successfully withstand the difficulties offered by coastal climates by incorporating these strategies, providing both safety and sustainability in the face of environmental obstacles.

19.2.3: **Ecosystem** **Integration:** Integrating ecosystems is crucial to coastal architecture. Implementing raised pathways and boardwalks lessens disturbance to animals and helps prevent damage to the delicate root systems of mangroves. Utilizing floating buildings, which can adjust to changing water levels and ensure minimum influence on the surrounding ecosystem, is another creative strategy. These tactics improve the overall sustainability and resilience of coastal projects while also safeguarding delicate ecosystems.

19.3: CASE STUDIES IN SUSTAINABLE ISLAND DWELLING DESIGN

19.3.1: Sustainable projects in Tuvalu: Tuvalu, a nation grappling with the adverse impacts of increasing sea levels, prioritizes the implementation of sustainable housing projects. In Tuvalu, there are ongoing initiatives that focus on the development of robust and ecologically sustainable infrastructure, with the objective of mitigating the adverse effects of climate change. These attempts include novel strategies in the field of building, including the implementation of elevated designs and the use of renewable energy sources. Tuvalu needs to establish a sustainable future for its island residents via the implementation of these programs.

19.3.2: Soneva Fushi, Maldives: Soneva Fushi is a model of luxury and sustainability. It is located in the Maldives. This upscale resort features waste-to-energy systems, eco-friendly villas, and a comprehensive coral propagation program. It not only provides sumptuous lodging but also acts as a role model for eco-friendly island living.

19.3.3: Bali's Bamboo Village: Bamboo Village, a model eco-resort in Bali, highlights the adaptability and sustainability of bamboo in building. This resort uses cutting-edge architectural strategies to take advantage of bamboo's regenerative qualities and provide an environment that blends well with its natural surroundings. Techniques for passive cooling highlight the resort's dedication to environmental responsibility.

19.3.4: Cambodia's Song Saa Private Island: Song Saa Private Island, which is located off the coast of Cambodia, is a pioneer in combining luxury and marine conservation. Overwater villas at the resort have breath-taking vistas and follow sustainable design principles. Song Saa is a great example of sustainable island living, with cutting-edge wastewater treatment technologies and a dedicated marine reserve demonstrating its genuine dedication to protecting the delicate marine habitat[140].

19.4: CHALLENGES AND CONSIDERATIONS

For projects on mangrove islands, sustainable design has great potential, but there are several important obstacles to overcome[141]:

19.4.1. Cost: Cost management is crucial to sustainable design. Sustainable qualities often result in significant long-term savings, though sometimes requiring greater initial costs. It is crucial to determine if the project is economically feasible and to make sure that the sustainable design you have selected is compatible with the project's overall profitability. This strategy guarantees that sustainability activities are both economically and ecologically sound.

19.4.2: Local Adaptation: Effective sustainable design on mangrove islands depends on local adaptability. Each island has distinct ecosystems and

---

[140] Musallam, A. (2019). Sustainable architecture in Saudi Arabia: Current challenges and future opportunities. Sustainability, 11(3), 752.

[141] Yang, X., & Ma, W. (2019). Sustainable architecture and development in China. Routledge.

needs, necessitating customized solutions. Therefore, it's crucial to use site-specific designs that take into account the island's unique qualities. Additionally, it is essential to include regional communities in the design process. Their understanding of culture and the surroundings may provide priceless insights, ensuring that the design is in harmony with the requirements and values of the community. This participatory method promotes sustainability over the long run by encouraging a feeling of ownership and responsibility over the planned areas.

19.4.3: **Regulatory Compliance:** When developing a sustainable design for mangrove islands, adherence to legal standards is crucial. This requires effectively negotiating the complex web of licenses and rules required for coastal development. Furthermore, it is crucial to carry out thorough Environmental Impact Assessments (EIAs). These evaluations provide a thorough knowledge of how planned buildings can possibly impact the sensitive environment of the mangroves. It is possible to create sustainable designs in a way that respects and preserves the environment by ensuring regulatory compliance and completing thorough EIAs. This strategy promotes responsible growth while preserving the ecosystem's integrity.

19.5: CONCLUSION

The development of mangrove island projects successfully depends on sustainable architecture. Architects and developers may build buildings that not only survive the rigors of coastal regions but also help to preserve delicate ecosystems by integrating eco-friendly design concepts and cutting-edge technology. These initiatives, if carefully considered, may show how sustainable living and environmental preservation can coexist, providing a global example for sustainable island habitation.

Chapter 20:

INCORPORATING MANGROVES INTO DWELLING DESIGN

20.1: INCORPORATING MANGROVES INTO DWELLING DESIGN

Mangroves may improve both the environmental sustainability and the aesthetic appeal of buildings when it comes to the construction of sustainable island homes. The numerous ways that mangroves may be effortlessly incorporated into home design for mangrove island developments are explored in this chapter. We'll analyze the advantages, difficulties, and creative solutions related to this strategy using cases from throughout the globe.

20.2: THE ECOLOGICAL SIGNIFICANCE OF MANGROVES

Before getting into mangrove incorporation into home design, it's important to understand their biological significance[142]:

20.2.1: **Coastal Protection:** Coastal protection is a critical function of mangrove ecosystems, providing vital benefits to both natural and human surroundings. Mangroves operate as natural barriers, successfully protecting coastal communities from the disastrous effects of storm surges and erosion. Their complex root systems are critical in wave attenuation, decreasing the power of incoming waves during high tides and possible tsunamis. Mangroves serve an important role in increasing the resilience of coastal communities and protecting the integrity of essential infrastructure along shorelines via these processes.

20.2.2: **Biodiversity Hotspots:** Mangrove environments are well-known biodiversity hotspots that provide essential support to a diverse range of organisms. Mangroves perform an essential function as nurseries in the life cycles of many marine species, which eventually supports local fisheries. Additionally, a variety of bird species depend on these environments as vital foraging and breeding grounds. This significant contribution to regional biodiversity emphasizes the need of protecting and restoring mangrove forests for the overall health and vitality of coastal ecosystems.

---

[142] Alongi, D. M. (2008). Mangrove forests: Resilience, protection from tsunamis, and responses to global climate change. *Estuarine, Coastal and Shelf Science*, 76(1), 1-13.

20.2.3: Carbon Sequestration: When it comes to properly absorbing and storing significant quantities of carbon in their biomass and soil, mangrove ecosystems play a key role in carbon sequestration. Along with seagrasses and salt marshes, mangroves fall under the category of "blue carbon" ecosystems because of their ability to help mitigate climate change. Mangroves are crucial in attempts to fight climate change and maintain coastal habitats because of the significant role they play in sequestering carbon.

20.3: INTEGRATING MANGROVES INTO DWELLING DESIGN

Mangroves must be considered in the design of homes, and this demands a multidisciplinary approach. Here are some tactics and things to think about[143]:

20.3.1: Elevated Walkways and Viewing Decks: Elevated pathways and observation platforms are implemented to provide locals and tourists with an immersive experience while having a minimum influence on the sensitive mangrove habitat. These buildings provide a platform for educational

---

[143] Primavera, J. H., & Esteban, J. M. A. (2008). A review of mangrove rehabilitation in the Philippines: successes, failures, and future prospects. *Wetlands Ecology and Management*, 16(5), 345-358.

possibilities, which improves people's awareness of and comprehension of the value of mangrove ecosystems. Incorporating interpretative signs also promotes a closer relationship between people and nature by serving as an educational tool to spread knowledge about the ecological importance of mangroves.

20.3.2: Mangrove Buffer Zones: In order to protect these delicate ecosystems, it is crucial to enforce setback restrictions in order to provide buffer zones between buildings and mangrove forests. A further step to strengthen the preservation of mangroves is to include native plants in these buffer zones. This tactical approach supports not just the sustainability and resilience of coastal communities overall but also the preservation of the integrity of mangrove ecosystems.

20.3.3: Floating Structures: Utilizing cutting-edge design ideas, such floating residences and structures that adjust to tidal fluctuations, demonstrates a forward-thinking strategy to limit harm to mangroves. By putting sustainability first, these buildings may include solar panels and rainwater collecting systems, further increasing their environmental impact. This all-encompassing strategy not only blends in with the surroundings but also establishes a model for resilient coastal life.

20.3.4: Mangrove Restoration: Mangrove restoration activities may be strengthened by strategic relationships with local conservation groups, incorporating them neatly into the development plan. Employing native mangrove species in landscaping not only increases the visual appeal but also produces sustainable gardens that blend with the surrounding environment. This technique not only helps to the restoration of essential coastal habitats but also generates regions of natural beauty.

20.4: REAL-WORLD EXAMPLES

20.4.1: **Mangrove-Focused** **Sustainability Initiatives in Tuvalu:** Innovative initiatives in Tuvalu show how human development and mangrove preservation can coexist peacefully. In addition to preserving marine environments, including essential mangrove habitats, the Tuvalu Funafuti Marine Conservation Area (TCMA) also engages local communities through educational programs, raising awareness of their crucial function in coastal ecosystems. Similar to this, the Vaiaku Lagi Hotel in Funafuti is an example of sustainable tourism because it uses energy-saving techniques and eco-friendly practices to lessen its impact on the nearby mangrove ecosystems. The Tuvalu National Library and Archives is another example of thoughtful architectural design that gives the nearby mangroves top priority. Beyond its architectural importance, it functions as a center for culture and education, highlighting the necessity of preserving regional ecosystems, including mangroves, for future generations.

20.4.2: Green School Bali, Indonesia: The Green School in Bali, Indonesia, is a groundbreaking example of how to integrate nature into architecture. The campus skillfully blends the natural world and education by incorporating bamboo groves and a lush rainforest into its layout. As part of its commitment to sustainability, the school uses eco-friendly building materials, renewable energy sources, and ecological landscaping, establishing a significant new standard for comprehensive environmental education.

20.4.3: Mangrove Village, The Philippines: The Philippines' Mangrove Village project, which incorporates essential mangrove forests into its design, is an example of holistic community development. Building elevated homes on stilts helps the mangroves below to flourish as well as protect against flooding. This novel approach demonstrates a peaceful coexistence of human habitation and nature by offering not only a sustainable living environment but also by nurturing the delicate ecosystem.

20.4.4: Koh Tao Bamboo Huts, Thailand: The Koh Tao Bamboo Huts in Thailand are an example of eco-friendly construction. Mostly made of bamboo, they seamlessly blend in with the nearby mangrove ecosystem. These huts prioritize the preservation of delicate mangrove roots by using an elevated design. This considerate approach highlights the value of coexisting with and protecting natural habitats while also showcasing sustainable building techniques[144].

20.5: CHALLENGES AND CONSIDERATIONS

While incorporating mangroves into house design provides various advantages, challenges must be addressed:

20.5.1: **Regulatory Compliance:** When commencing on mangrove integration initiatives, regulatory compliance is crucial. This includes negotiating rigorous regulatory systems and conforming to local restrictions about coastal development and mangrove protection. Additionally, undertaking comprehensive Environmental Impact Assessments is vital. These analyses aim to assure that the integration activities are in perfect agreement with larger ecological preservation goals, guaranteeing a sustainable and balanced coexistence.

---

[144] Dahdouh-Guebas, F., Jayatissa, L. P., Di Nitto, D., Bosire, J. O., Lo Seen, D., & Koedam, N. (2005). How effective were mangroves as a defence against the recent tsunami? *Current Biology*, 15(12), R443-R447.

20.5.2: Maintenance and Monitoring: To sustain the integrity of mangrove integration, a comprehensive maintenance and monitoring structure is required. This covers the application of sustainable maintenance procedures, which are crucial in preserving the health and vitality of mangrove regions. Moreover, frequent monitoring practices are needed. These monthly examinations act as an early warning system, allowing the prompt identification of any indicators of stress or deterioration within the mangrove ecosystem. This proactive strategy assures the continuous success of the integration initiatives.

20.5.3: Community Engagement: The cornerstone of successful mangrove integration initiatives is community involvement. It is possible to promote a feeling of ownership and stewardship by actively integrating neighborhood groups in the design and maintenance stages. This cooperative strategy increases the likelihood of long-term success by ensuring that the integration is in line with the needs and values of the community. In addition, educational programs are essential in spreading the word about the fundamental importance of mangrove ecosystems. It is possible to strengthen the sustainability of mangrove-integrated structures by fostering a culture of respect and conservation among locals and tourists[145].

---

[145] Donato, D. C., Kauffman, J. B., Murdiyarso, D., Kurnianto, S.,

20.6: CONCLUSION

A comprehensive strategy for environmentally friendly architecture is to include mangroves into the construction of homes for mangrove island developments. Architects and developers may build structures that not only live peacefully with nature but also help to preserve these priceless coastal ecosystems by acknowledging the biological importance of mangroves and strategically incorporating them into design plans. The incorporation of mangroves into building design serves as a paradigm for sustainable island life around the globe thanks to creative solutions and community engagement.

Stidham, M., & Kanninen, M. (2011). Mangroves among the most carbon-rich forests in the tropics. *Nature Geoscience*, 4(5), 293-297.

Chapter 21:

INNOVATIVE BUILDING MATERIALS AND TECHNIQUES

For mangrove island developments, sustainable house design relies heavily on cutting-edge materials and methods of construction. In addition to lowering the environmental impact of construction, these materials and techniques also help make the buildings resilient and sustainable in the long run. This chapter will investigate modern construction methods and materials that may be used into architectural plans for mangrove islands, taking inspiration from all across the globe.

21.1: SUSTAINABLE BUILDING MATERIALS

21.1.1: Bamboo: Due to its quick growth and renewability, bamboo has become well-known as a sustainable construction material. It is useful for a variety of architectural applications because to its great strength, toughness, and adaptability.

Examples: The Green School in Bali, Indonesia, demonstrates the structural versatility and aesthetic attractiveness of bamboo construction[146].

21.1.2: Recycled and reused resources: Reusing and recycling resources, including salvaged steel, reused wood, and repurposed shipping containers, may considerably lessen the environmental effect of construction.

Examples: The Beach Box in Amagansett, New York, is a beachside resort built out of repurposed shipping containers[147].

21.1.3: Rammed Earth: To build walls, rammed earth construction requires compacting natural raw materials like gravel, earth, chalk, or lime inside of forms. It is a low-energy, environmentally friendly method with superior thermal capabilities.

Example: Rammed earth construction include the Great Wall of WA in Australia, which exemplifies its sturdiness and strength[148].

---

[146] Ghavami, K. (2005). Bamboo as reinforcement in structural concrete elements. Cement and Concrete Composites, 27(6), 637-649.

[147] Gorgolewski, M. (2016). Reclaimed building materials as an element of sustainable urban design. Journal of Urban Design, 21(2), 246-263.

[148] Hall, M. R., & Reed, M. S. (2015). A review of the social and cultural factors influencing sustainable rammed earth construction. Building and Environment, 89, 243-250.

21.1.4: Straw Bales: Straw bales are excellent insulators and are used as construction components. They have a little carbon impact and are renewable and biodegradable.

Example: As an illustration of the viability of straw bale building, consider The BaleHaus at the University of Bath, a sustainable housing project[149].

21.1.5: Green Concrete: Technological advancements in the manufacture of concrete have resulted in the creation of ecologically friendly choices, such as concrete with a high fly ash content or produced from recycled aggregates.

Example: the construction of Amsterdam's The Edge, the so-called greenest skyscraper in the world, made use of sustainable concrete[150].

21.2: MODERN CONSTRUCTION METHODS

21.2.1: 3D printing: By enabling the construction industry to produce complex buildings with little waste, 3D printing technology is transforming the industry. Using this method, you may create

---

[149] Walker, P. (2012). Design and construction of straw bale buildings. Building and Environment, 49, 1-8.

[150] Tam, V. W. Y., Tam, C. M., & Le, K. N. (2007). Environmental performance of concrete made with crushed glass as partial replacement of natural aggregates. Resources, Conservation and Recycling, 52(5), 747-757.

buildings that are both visually beautiful and environmentally responsible. As an example, the building business Apis Cor, which employs 3D printing, has shown that it is possible to create dwellings[151].

21.2.2: Prefabrication and Modular Construction: To cut down on waste and construction time, prefabrication entails producing building components in a factory and putting them together on-site. An innovative modular construction with mobility and sustainability for coastal life is the Arkup Floating Home[152].

21.2.3: Passive home Design: Passive home design focuses on developing very energy-efficient buildings with low heating and cooling demands. These designs make use of heat recovery technologies, airtight structure, and insulation. As an example, the Bahnstadt District in Heidelberg, Germany, uses passive home design concepts to cut down on emissions and energy usage[153].

---

[151] Le, T. A., Austin, S. A., Lim, S., Buswell, R. A., & Gibb, A. G. (2012). Development of an automated concrete printing system for building components. Rapid Prototyping Journal, 18(2), 129-143.
[152] Smith, R., & Tran, H. (2018). An investigation into the benefits of offsite prefabrication in reducing construction waste. Sustainability, 10(7), 2475.
[153] Feist, W., Schnieders, J., & Dorer, V. (2005). International Passive House standard—Design criteria and main performance indicators. Passive House Institute, Darmstadt, Germany.

21.3: SUSTAINABLE ISLAND DWELLING DESIGN

It takes considerable planning and attention to incorporate these cutting-edge construction materials and methods into sustainable island living designs. Here are some important things to remember:

21.3.1: Local Context: When adapting construction methods to the particular features of the island, it is important to examine factors such as climate, weather patterns, and resource availability.

21.3.2: Resilience: Design and construct resilient structures capable of withstanding the challenges posed by coastal environments, including rising sea levels, saltwater intrusion, and extreme weather phenomena.

21.3.3: Community Engagement: Ensure that projects are in line with local communities' wants and requirements by including them in the planning and construction process.

21.3.4: Long-Term Sustainability: Give long-term sustainability top priority by taking into account the material lifespan, upkeep needs, and energy efficiency.

21.3.5: Regulatory Compliance: To assure compliance and project profitability, navigate regional and global legislation pertaining to building and sustainability.

21.4: REAL-WORLD EXAMPLES

21.4.1: The Arkup Floating Home, Miami, USA: The Arkup is a solar-powered, hurricane-resistant floating house that blends cutting-edge building methods with sustainability. It is located in Miami, United States. In order to gain Sustainability, the home uses rooftop solar panels to produce its electricity and rainwater collection to ensure self-sufficiency[154].

21.4.2: Amsterdam, Netherlands' The Edge: Green Building: The Edge is an intelligent, environmentally friendly office building that makes use of sustainable concrete and energy-saving design ideas. For a better working environment, it includes cutting-edge air quality monitoring and creative space usage.

21.4.3: The Great Wall of WA, Australia: Rammed Earth tourist destination includes walls built using rammed earth methods, demonstrating the durability and aesthetic appeal of sustainable materials[155].

---

[154] The Arkup Floating Home (Official Website)

[155] The Edge, Amsterdam, Netherlands (Official Website)

21.5: CHALLENGES AND CONSIDERATIONS

While using cutting-edge construction materials and methods has many advantages, there are certain obstacles to overcome:

21.5.1: Cost: Some innovative building techniques and environmentally friendly materials may have a greater initial cost, necessitating a long-term investment viewpoint.

21.5.2: Availability: The accessibility of sustainable resources might differ from place to place, calling for careful supply chain management and procurement.

21.5.3: Skilled Labor: The need for skilled labor may be necessary for specialized building procedures, which may affect the viability and cost of a project.

21.5.4: Regulatory Obstacles: Advocacy and adaptation are often necessary since building laws and regulations often need to permit the use of cutting-edge materials and methods[156].

21.6: CONCLUSION

To reduce the negative environmental effects of coastal expansion and increase resilience in the face

---

[156] The Great Wall of WA, Australia (Great Wall of WA)

of climate change, sustainable island home designs must use cutting-edge building materials and construction methods. To design ecologically sound buildings that complement the distinct beauty and difficulties of mangrove island environments, architects, builders, and communities must collaborate. We can strike the delicate balance between human occupancy and the preservation of environment on these biodiverse coastal areas by embracing innovation.

PART VIII: COMMUNITY ENGAGEMENT AND CONSERVATION

Chapter 22:

INVOLVING LOCAL COMMUNITIES IN MANGROVE RESTORATION

Mangrove restoration is an important activity for sustaining the health of coastal ecosystems and assuring the sustainability of these essential habitats. While the technical parts of restoration, such as choosing appropriate species and planting procedures, are crucial, the success of restoration initiatives typically rests on the cooperation of local people. In this chapter, we will discuss the necessity of involving local people in mangrove restoration initiatives, the advantages it offers, and ways for successful community engagement. We will also look into case studies and examples that illustrate successful community-driven mangrove restoration programs from throughout the globe.

22.1: THE SIGNIFICANCE OF COMMUNITY ENGAGEMENT IN MANGROVE RESTORATION

22.1.1: Local Knowledge and Expertise: Local people contain unique information about their ecosystems, including the historical distribution of mangroves, traditional usage of mangrove resources, and local biological dynamics. This information may considerably inform restoration planning and implementation[157].

22.1.2: Ownership and Stewardship: When people actively engage in restoration initiatives, they create a feeling of ownership and responsibility for the newly restored mangrove ecosystems. This stewardship approach may lead to long-term preservation and upkeep.

22.1.3: Sustainability and Livelihoods: Many coastal people rely on mangrove ecosystems for their livelihoods, including fishing and tourism. Engaging these communities in rehabilitation helps assure the sustainability of their livelihoods.

---

[157] Dahdouh-Guebas, F., et al. (2005). How effective were mangroves as a defense against the recent tsunami? Current Biology, 15(12), R443-R447.

22.2: STRATEGIES FOR EFFECTIVE COMMUNITY INVOLVEMENT

Any mangrove restoration project's effectiveness depends on including the local community in conservation efforts. In this section, we examine several tactics for encouraging active community involvement and participation in the restoration process[158].

22.2.1: Participatory Planning: Involve local community members in the planning process from the beginning. Seek their input on project goals, methods, and potential challenges.

22.2.2: Capacity Building: Offer community members chances for training and capacity building, including instruction in mangrove ecology, restoration strategies, and monitoring techniques.

22.2.3: Collaborative Decision-Making: Encourage cooperation between researchers, NGOs, government organizations, and local communities. Shared decision-making can result in initiatives that are more inclusive and productive.

22.2.4: Economic Incentives: Consider developing financial incentives for communities to take part in restoration efforts. Activities including

[158] Lewis III, R. R. (2005). Ecological engineering for successful management and restoration of mangrove forests. Ecological Engineering, 24(4), 403-418.

ecotourism or the sustainable exploitation of mangrove products are two examples.

22.3: CASE STUDIES IN COMMUNITY-DRIVEN MANGROVE RESTORATION

22.3.1: Bako, Malaysia: The mangrove restoration program in Malaysia's Bako National Park is a prime illustration of how thoughtful conservation efforts may result in several advantages. This initiative not only dramatically improved the park's ecological health, but it also served as an attraction for ecotourists who came to see the flourishing mangrove environment in its restored grandeur.

22.3.2: Krishnapuram, India: The effective restoration of mangroves in Krishnapuram, India, is evidence of the practical advantages such initiatives may have. This program has reduced coastal erosion while strengthening the region's resistance to storms by reviving the mangrove environment, offering a crucial barrier for the coastal communities[159].

22.3.3: Tumbes, Peru: A unique mangrove regeneration project in the coastal area of Tumbes, Peru, has produced notable benefits. This program, spearheaded by committed local fishing clubs, has

---

[159] Alongi, D. M. (2008). Mangrove forests: Resilience, protection from tsunamis, and responses to global climate change. Estuarine, Coastal and Shelf Science, 76(1), 1-13.

seen a considerable increase in fish numbers and a general resurgence of the ecology. Such initiatives highlight the critical function that community-led restoration initiatives have in reviving coastal habitats and maintaining livelihoods[160].

22.4: CHALLENGES AND CONSIDERATIONS

22.4.1: Land Tenure and Rights: In order to avoid disagreements and make sure that local people get the most out of restoration efforts, it is essential to find solutions to concerns relating to land tenure and property rights. This action establishes the framework for effective and long-lasting mangrove restoration initiatives[161].

22.4.2: Climate Change Resilience: In order to combat the increasing risks brought on by sea level rise and extreme weather events, climate change resilience techniques must be included into restoration programs. In the face of changing environmental difficulties, this strategic strategy guarantees the long-term viability and sustainability of mangrove restoration programs[162].

---

[160] Alongi, D. M. (2008). Mangrove forests: Resilience, protection from tsunamis, and responses to global climate change. Estuarine, Coastal and Shelf Science, 76(1), 1-13.

[161] Ellison, A. M., & Farnsworth, E. J. (1996). Anthropogenic disturbance of Caribbean mangrove ecosystems: past impacts, present trends, and future predictions. Biotropica, 328-334.

22.4.3: Long-Term Sustainability: Community-led restoration initiatives depend on ongoing upkeep, attentive monitoring, and deliberate modifications to ensure their long-term viability. These steps are necessary for maintaining thriving mangrove ecosystems throughout time, enabling them to flourish and provide crucial advantages to the environment and nearby populations[163].

22.5: CONCLUSION

To achieve ecological, social, and economic sustainability, local community participation in mangrove restoration is not only a practical need but also a potent tool. Communities get several advantages by engaging in mangrove conservation and management, including cleaner water, better jobs, and a closer connection with nature. Community involvement is still crucial to the success of mangrove restoration efforts notwithstanding our progress in combating coastal degradation and climate change.

---

[162] Giri, C., et al. (2011). Status and distribution of mangrove forests of the world using earth observation satellite data. Global Ecology and Biogeography, 20(1), 154-159.

[163] Primavera, J. H., & Esteban, J. M. A. (2008). A review of mangrove rehabilitation in the Philippines: successes, failures, and future prospects. Wetlands Ecology and Management, 16(5), 345-358.

Chapter 23:

EDUCATION AND AWARENESS INITIATIVES IN MANGROVE RESTORATION

Restoring mangrove ecosystems is necessary because of the vital role they play in coastal resiliency and biodiversity conservation. However, education and awareness programs are just as important to the success of mangrove restoration projects as technical knowledge and community engagement. In this chapter, we will discuss the role that public awareness and education initiatives have in the process of mangrove restoration. We will go through the goals, methods, and outcomes of such efforts, as well as provide examples of effective curriculum from across the world.

23.1: THE IMPORTANCE OF RAISING PUBLIC KNOWLEDGE IN RESTORING MANGROVES

Comprehending the intricacies of mangrove ecosystems is of paramount importance in the context of restoration endeavors, as it facilitates the promotion of active engagement and enduring conservation by means of community education on their indispensable contributions to erosion mitigation and biodiversity conservation. This understanding serves as the cornerstone for many effective initiatives[164]:

Raising Awareness: Communities and stakeholders may be better educated about the value of mangrove ecosystems, the risks they face, and the rewards of restoration via education and awareness campaigns. These campaigns may reach locals, politicians, visitors, and the world at large.

Building Support: People and groups that have more information are more inclined to fund and take part in mangrove restoration projects. In order to ensure the long-term success of a project, public support is frequently essential.

Changing Behavior: Education activities may help reduce human impacts on mangroves by encouraging more sustainable lifestyle choices. Excessive extraction of materials, air and water pollution, and poor use of land are all examples.

Capacity Building: Communities may take an active role in restoration operations like planting and monitoring with the help of training and education initiatives.

---

[164] Dahdouh-Guebas, F., et al. (2005). How effective were mangroves as a defense against the recent tsunami? Current Biology, 15(12), R443-R447.

23.2: OBJECTIVES OF EDUCATION AND AWARENESS INITIATIVES

The following are the goals of projects to raise awareness and educate people about mangrove restoration[165]:

Informing about Mangrove Ecology: Participant learning objectives should include an understanding of mangroves' ecological value as carbon sinks, coastline protectors, and primary marine life nurseries.

Highlighting Threats: Mangroves need more people to know about the dangers they face from things like deforestation, habitat loss, pollution, and the effects of climate change.

Promoting Sustainable Practices: Support initiatives that help people survive off the land without damaging it, such as ecotourism, sustainable fishing, and careful resource management.

Advocating Policy Changes: Rally public and political support for measures at the regional, national, and international levels that will conserve and restore mangrove ecosystems.

---

[165] Lewis III, R. R. (2005). Ecological engineering for successful management and restoration of mangrove forests. Ecological Engineering, 24(4), 403-418.

23.3: STRATEGIES FOR EFFECTIVE EDUCATION AND AWARENESS INITIATIVES

Community Involvement: Participation from local communities in program design and execution increases the likelihood of success in meeting cultural needs.

Multimedia Campaigns: Reach out to people where they are by using print, radio, television, and online social networks.

Hands-on Workshops: Facilitate learning by holding seminars, taking students on field excursions, and giving them real-world experiences.

School Programs: Work with educational institutions to include lessons on mangroves in classroom and after-school programs.

Partnerships: The best way to make use of existing resources and knowledge is to join forces with other environmental groups, NGOs, government agencies, and the commercial sector[166].

---

[166] Primavera, J. H., & Esteban, J. M. A. (2008). A review of mangrove rehabilitation in the Philippines: successes, failures, and future prospects. Wetlands Ecology and Management, 16(5), 345-358.

23.4: CASE STUDIES IN SUCCESSFUL EDUCATION AND AWARENESS INITIATIVES

There are several case studies that illustrate successful education and awareness initiatives. In this context, we may see the following examples[167]:

23.4.1: Mangroves for the Future (MFF): The MFF is an Asian project that uses media and educational activities to advocate for responsible mangrove management and restoration. Through these channels, they work to raise public consciousness.

23.4.2: Mangrove Action Project (MAP): MAP works with locals in many different nations to protect mangroves by providing educational opportunities including seminars and training. They take a comprehensive strategy by providing resources for learning, advocacy, and economic security.

23.4.3: Jamaica's Alligator Head Foundation: This Jamaican group protects mangroves and coral reefs with a multi-pronged approach that includes teaching, scientific inquiry, and outreach to local communities. Their activities range from research to instruction for local fishermen[168].

---

[167] Alongi, D. M. (2008). Mangrove forests: Resilience, protection from tsunamis, and responses to global climate change. Estuarine, Coastal and Shelf Science, 76(1), 1-13.

23.5: CHALLENGES AND CONSIDERATIONS

Cultural Sensitivity: The success and cultural appropriateness of educational programs depends on their respect for and incorporation of local knowledge, beliefs, and customs.

Resource Constraints: Lack of sufficient financing and resources might limit the effectiveness and reach of educational campaigns. It may be required to work with outside partners and funders.

Monitoring and Evaluation: Measure the efficacy of your education and awareness programs on a regular basis so you may tweak them as needed[169].

23.6: CONCLUSION

The revitalization and preservation of mangrove ecosystems rely heavily on education and awareness campaigns. These initiatives encourage a feeling of stewardship, spread awareness of sustainable habits, and push for legislative changes that are good for people and the planet. As mangrove restoration

[168] Giri, C., et al. (2011). Status and distribution of mangrove forests of the world using earth observation satellite data. Global Ecology and Biogeography, 20(1), 154-159.
[169] Ellison, A. M., & Farnsworth, E. J. (1996). Anthropogenic disturbance of Caribbean mangrove ecosystems: past impacts, present trends, and future predictions. Biotropica, 328-334.

efforts continue to expand worldwide, effective education and awareness campaigns will be essential in ensuring the long-term health and resilience of these critical coastal habitats.

Chapter 24:

LEGAL FRAMEWORKS AND PROTECTION OF MANGROVE ECOSYSTEMS

Coastal people and wildlife rely on mangrove ecosystems, yet these ecosystems are under danger from things like deforestation, pollution, and habitat loss. Protective laws and institutional safeguards are crucial for the long-term survival of mangrove forests. This chapter will examine the legal framework for protecting mangroves, touching on international treaties, domestic statutes, and local ordinances. We'll take a look at how these laws and treaties protect mangroves and advance eco-friendly policies.

24.1: THE VALUE OF LAWS AND REGULATIONS IN PROTECTING MANGROVES

The preservation and management of mangroves are effectively facilitated by the implementation of laws and regulations, which serve as essential mechanisms for establishing legal frameworks. These frameworks are vital in ensuring the protection and sustainable management of mangrove ecosystems. The significance of these entities resides in[170]:

Global Significance: The protection of mangrove ecosystems is a worldwide issue because of the many advantages they bring on ecological, economic, and social levels. Addressing transboundary concerns and fostering international collaboration are both aided by legal frameworks.

Protection from Deforestation: For reasons including agriculture, aquaculture, and urban growth, it is vital to safeguard mangrove forests from illegal destruction.

Safeguarding Biodiversity: Mangroves are home to a wide variety of rare and vulnerable species. The establishment of legal frameworks is crucial to the preservation of these species and their environments.

Community Rights: It is crucial for legal frameworks to recognize and safeguard the rights of local populations that rely on mangroves for subsistence.

---

[170] Spalding, M., Kainuma, M., & Collins, L. (2010). World Atlas of Mangroves. Earthscan.

24.2: INTERNATIONAL AGREEMENTS FOR MANGROVE CONSERVATION

The importance of international agreements cannot be overstated in terms of facilitating the coordination of worldwide initiatives aimed at the preservation and conservation of mangroves. Several significant agreements have been identified, namely[171]:

Ramsar Convention: The Ramsar Convention, an international treaty focused on the conservation and sustainable use of wetlands, acknowledges the significant ecological value of mangroves as crucial components of wetland ecosystems. Nations that have ratified the treaty are obligated to uphold the responsibility of safeguarding and preserving these regions.

Convention on Biological Diversity (CBD): The protection of biodiversity, including mangrove habitats, is actively supported by the CBD. The aforementioned statement promotes the adoption of strategies by relevant stakeholders to ensure the sustainable use and conservation of mangrove ecosystems.

United Nations Framework Convention on Climate Change (UNFCCC): Mangroves are well

---

[171] Ellison, A. M., & Farnsworth, E. J. (1996). Anthropogenic disturbance of Caribbean mangrove ecosystems: past impacts, present trends, and future predictions. Biotropica, 328-334.

acknowledged for their significant contribution to both climate change mitigation and adaptation efforts. The United Nations Framework Convention on Climate Change (UNFCCC) provides backing for initiatives aimed at mitigating greenhouse gas emissions resulting from deforestation and forest degradation, especially in mangrove ecosystems.

24.3: NATIONAL LEGAL FRAMEWORKS FOR MANGROVE PROTECTION

National laws protect mangrove habitats. Some common elements of national mangrove preservation laws are[172]:

Mangrove Forest Protection Laws: Numerous nations have enacted distinct legal frameworks and regulatory measures with the primary objective of safeguarding mangrove forests. Frequently, these regulations serve to restrict the engagement in illicit logging, land conversion, and several other activities that are seen to be detrimental to the environment.

Environmental Impact Assessment (EIA): In certain legal contexts, developers are often mandated to carry out Environmental Impact Assessments

[172] McLeod, E., et al. (2010). A blueprint for blue carbon: toward an improved understanding of the role of vegetated coastal habitats in sequestering CO2. Frontiers in Ecology and the Environment, 9(10), 552-560.

(EIAs) as a prerequisite for commencing projects in close proximity to mangrove habitats. Environmental Impact Assessments (EIAs) play a crucial role in evaluating and analyzing the possible environmental consequences of a proposed project or activity. They also serve to identify and recommend appropriate ways to mitigate these effects.

Marine Protected Areas (MPAs): The incorporation of mangrove ecosystems inside Marine Protected Areas (MPAs) is an often used legal approach for the aim of conservation. These locations often use specific measures designed to protect and preserve biodiversity.

Community-Based Regulations: In some regions, individuals living in the vicinity actively participate in the management and preservation of mangrove ecosystems, taking a significant role in these collective efforts. The recognition and support of community-based conservation initiatives are appropriately recognised and promoted within legislative frameworks.

24.4: CHALLENGES IN IMPLEMENTING LEGAL FRAMEWORKS

Enforcement: The inadequate enforcement of rules and regulations is a significant challenge in several countries. This phenomena may occur as a result of limitations in the resources that are accessible, occurrences of corrupt practices, or conflicting interests.

Lack of Awareness: The limited knowledge and understanding shown by some organizations and stakeholders about the already established legal protections for mangroves might potentially hinder their compliance with these regulatory measures.

Land Tenure Issues: Conflicts stemming from land ownership problems might complicate mangrove conservation programs. Legal systems must be able to handle the complex problems of land rights and ownership disputes in an efficient manner[173].

24.5 CASE STUDIES IN EFFECTIVE LEGAL FRAMEWORKS

Here are example studies of efficient mangrove protection laws:

Thailand's Mangrove Forest Conservation Act: To protect its mangrove forests, Thailand has enacted a number of laws, one of which is the Mangrove Forest Conservation Act. This law protects mangrove forests by requiring Environmental Impact Assessments (EIAs) before

---

[173] Pomeroy, R. S., & Douvere, F. (2008). The engagement of stakeholders in the marine spatial planning process. Marine Policy, 32(5), 816-822.

any construction may take place in the immediate vicinity of these ecosystems.

Bonaire's Marine Park Regulations: Bonaire's marine park regulations and other legislation have been successful in protecting the island's mangroves and coral reefs. The community's engagement and cooperation with local law enforcement is essential.

Ecuador's Sustainable Development Initiatives: In order to ensure that conservation efforts and local community needs are adequately balanced, Ecuador's legal framework promotes sustainable development in mangrove zones. The regulations and incentives set up to do this are crucial.

24.6: COMMUNITY INVOLVEMENT IN LEGAL FRAMEWORKS

Community engagement in legislative frameworks for mangrove conservation is essential for success and sustainability. Some major features of community engagement in legal frameworks:

Community-Based Conservation Agreements: Mangrove conservation projects may be more successful and long-lasting if local communities are included in policymaking and get fair compensation via legally binding agreements.

Indigenous Rights: It is of the highest significance that indigenous peoples' legal rights be recognized and upheld, since mangroves are often found in areas inhabited by indigenous peoples[174].

24.7: CONCLUSION

Mangrove ecosystems rely heavily on the creation and enforcement of legal frameworks and safeguards to ensure their survival. These measures contribute to the prevention of deforestation, the protection of biodiversity, and the promotion of sustainable practices. The resilience and long-term survival of mangroves depend significantly on the effective implementation and enforcement of conservation measures, as well as the active engagement of local communities and international collaboration.

[174] Barbier, E. B., et al. (2011). The value of estuarine and coastal ecosystem services. Ecological Monographs, 81(2), 169-193.

PART IX: SUCCESS STORIES AND CASE STUDIES

Chapter 25:

REAL-WORLD EXAMPLES OF MANGROVE-BASED ISLAND BUILDING

The use of mangroves into habitat restoration and architectural design has generated a surge of creativity in the search for sustainable island life. While the idea of cultivating mangrove trees for sustainable island habitations may seem innovative, actual instances from throughout the world show its viability and potential for revolutionary change. This chapter guides you through these exceptional initiatives while emphasizing their accomplishments, difficulties, and key takeaways.

25.1: THE GREEN BELT PROJECT, SAUDI ARABIA

Location: Al Khobar, Saudi Arabia Project Initiation Year: 1997

One of the most well-known examples of incorporating mangroves into urban development is the Green Belt Project in Al Khobar, Saudi Arabia.

The city launched a bold endeavor to establish a natural buffer against the expanding Arabian Gulf in response to the city's growing growth and environmental damage. Planting mangrove species along the shoreline helped achieve the project's goals of reducing erosion, increasing biodiversity, and providing locals with recreational spaces. In addition to its environmental benefits, the expanding mangrove forest has become a highly sought-after urban haven[175].

25.2: SUNKEN CITY, LOS ANGELES, USA

Location: San Pedro, Los Angeles, USA *Project Initiation Year*: Ongoing

In response to rising sea levels and the need for sustainable urban development, the Sunken City project in San Pedro, Los Angeles, is redefining what it means to live along the shore. Combining architectural originality with ecological restoration, this project makes use of mangroves in its design. Building a system of floating homes, boardwalks, and underwater habitats will not only regenerate coastal ecosystems but also provide resilient and sustainable housing options[176].

[175] Al-Naim, B. (2012). The Green Belt of Al Khobar: Urban development of mangrove areas in the Eastern Province of Saudi Arabia. Urban Biodiversity and Design. Springer.

[176] The Sunken City. (2009). https://www.thesunkencity.com.

25.3: GREEN CLIMATE FUND MANGROVES FOR FIJI

Location: Fiji Project Initiation Year: 2017

Fiji and other Pacific islands are in grave danger from rising sea levels. The Green Climate Fund's Mangroves for Fiji program is a shining example of the value of mangroves for both climate adaptation and mitigation. Establishing and maintaining mangrove forests along Fiji's beaches increases biodiversity, supports local livelihoods, and offers coastal protection. The project's success can be attributed to the all-encompassing approach that was taken in its design[177].

25.4: THE FLOATING UNIVERSITY, BANGLADESH

Location: Dhaka, Bangladesh Project Initiation Year: 2012

Bangladesh, a nation most susceptible to the adverse impacts of rising sea levels, has actively adopted inventive approaches to foster sustainable livelihoods. The Floating University located in Dhaka is an educational establishment that has been constructed exclusively on floating platforms, which have been seamlessly merged with the surrounding

---

[177] Green Climate Fund. (2014). Mangroves for Fiji
https://www.greenclimate.fund/projects/8264A965-2163-4D6B-88D2-66A42B282CC1.

mangrove environments. This innovative initiative showcases the potential of mangroves in fostering education and promoting sustainable urban development. The establishment not only offers a robust educational setting but also plays a role in safeguarding the waterways of Dhaka[178].

25.5: MOÍN MANGROVES, COSTA RICA

Location: Moín, Costa Rica Project Initiation Year: 2003

The Moín Mangroves restoration project in Costa Rica serves as a notable illustration of the efficacy of public-private cooperation in the realm of mangrove protection. In response to the detrimental effects of industrial operations, the primary objective of this endeavor was to undertake the restoration of more than 800 hectares of mangrove forests. By means of a synergistic partnership including governmental entities, local communities, and private firms, the project has successfully attained its ecological objectives while concurrently fostering economic prospects and facilitating communal development[179].

---

[178] ArchDaily. (2015). Floating University in Bangladesh Creates Floating Farming Community
https://www.archdaily.com/611106/floating-university-in-bangladesh-creates-floating-farming-community.

[179] Rovira, A., & Beck, S. G. (2010). Moín mangrove reforestation: a successful project. Global Partnership for the Restoration of Coastal

25.6: REWA ECO-LODGE, FIJI

Location: Rewa, Fiji Project Initiation Year: Ongoing

The Rewa Eco-Lodge in Fiji serves as a noteworthy exemplar of sustainable tourism, with a primary focus on the preservation and use of mangroves. The eco-friendly resort is situated in the midst of a mangrove forest, exemplifying the seamless integration of sustainable construction with the surrounding natural ecosystems. Mangroves are beautiful and have responsible tourism activities that support their conservation[180].

25.7: THE GREAT WALL OF LAGOS, NIGERIA

Location: Lagos, Nigeria Project Initiation Year: 2008

Nigeria's largest city, Lagos, has a critical problem: rapid urbanization has exacerbated the already serious problem of shoreline erosion. The Great Wall of Lagos project is an ambitious endeavor that combines the protection of the shoreline with the participation of the local community. This barrier, which is 8.7 kilometers in length, relies heavily on mangrove trees. This case study shows

and Marine Habitats.

[180] Rewa Eco-Lodge. (2007). Official Website https://www.rewaecolodge.com.

how new engineering approaches may be successfully integrated with natural systems to safeguard coastal areas at risk from environmental vulnerabilities[181].

25.8: SRI LANKAN MANGROVE RESTORATION PROJECT

Location: Sri Lanka Project Initiation Year: Ongoing

Sri Lanka's mangrove restoration project is an all-encompassing national effort to protect and revive mangrove habitats. The effort sets out to solve several problems at once by planting one million mangrove seedlings around the country. These problems include deforestation and pollution. This endeavor serves as a model for how to successfully undertake large-scale mangrove restoration projects, with positive results for the environment and society at large.Sri Lanka's mangrove restoration project is an all-encompassing national effort to protect and revive mangrove habitats. The program has the lofty objective of planting one million mangrove saplings around the country to combat challenges like deforestation and pollution. This program serves as a model for the large-scale restoration of mangroves, which has several environmental and social benefits,

[181] Lagos State Government. (2019). Lagos State Government Constructing World's Second Longest Wall https://lagosstate.gov.ng/blog/2019/10/16/lagos-state-government-constructing-worlds-second-longest-wall.

and which involves the participation of local people and the use of scientific knowledge[182].

25.9: THE PHILIPPINES SUSTAINABLE ECO-TOURISM DEVELOPMENT

Location: Palawan, Philippines Project Initiation Year: Ongoing

The initiative on Sustainable Eco-Tourism Development in Palawan, Philippines, showcases the capacity of mangroves to facilitate sustainable tourism. Through the implementation of measures aimed at safeguarding and rejuvenating mangrove ecosystems, this endeavor has successfully established an unparalleled eco-tourism encounter for anyone seeking to engage with nature. Tourists have the opportunity to engage in the exploration of mangrove forests, acquire knowledge about their biological significance, and see the positive outcomes experienced by local people as a result of engaging in responsible tourism practices[183].

---

[182] Asia Foundation. (2009). Sri Lanka: One Million Mangroves Restoration Project. https://asiafoundation.org/projects/sri-lanka-one-million-mangroves-restoration-project/

[183] The Philippines Sustainable Eco-Tourism Development. (2016). Official Website. https://www.psed.ph/

25.10: THE GREEN BELT OF GUJARAT, INDIA

Location: Gujarat, India Project Initiation Year: 2011

The Green Belt initiative implemented in Gujarat, India, serves as a notable illustration of efforts aimed at the restoration of coastal ecosystems with the dual purpose of mitigating erosion and fostering biodiversity. The implementation of mangrove plantation along the coastal areas has effectively reduced the adverse effects of rising sea levels and storm surges. Additionally, the project provides sustainable livelihood options to local populations via their active participation in mangrove restoration and conservation endeavors[184].

25.11: PANAMA BAY URBAN DEVELOPMENT PROJECT

Location: Panama City, Panama Project Initiation Year: Ongoing

The urban development project in Panama City situated in Panama Bay strategically integrates mangroves into its design as a means to effectively tackle environmental concerns. The objective of the

[184] India CSR Network. (2015). Mangrove Reforestation Project, Gujarat. https://indiacsr.in/mangrove-reforestation-project-gujarat/

project is to bolster the city's capacity to withstand the effects of increasing urbanization, namely by mitigating the impact of floods and enhancing the overall quality of water resources. This study demonstrates the integration of mangroves into urban development as a means to create an environmentally sustainable and habitable urban environment[185].

25.12: THE GREENING OF SINGAPORE

Location: Singapore Project Initiation Year: Ongoing

Singapore is dedicated to promoting environmentally sustainable urban development, which includes the restoration and conservation of mangrove forests. Initiatives like as the Mandai Mangrove and Mudflats restoration project exemplify the commitment of the city-state to ecological restoration and educational endeavors. These endeavors facilitate the conversion of previous industrial zones into flourishing mangrove ecosystems, therefore emphasizing the potential for urbanization and biodiversity preservation to live together[186].

---

[185] The Nature Conservancy. (2015). Panama Bay Urban Development Project. https://www.nature.org/en-us/what-we-do/our-insights/perspectives/panama-bay-urban-development-project/

25.13: THE MALDIVES COASTAL PROTECTION PROJECT

Location: Maldives Project Initiation Year: Ongoing

The Maldives, a country especially vulnerable to the impacts of rising sea levels, has initiated the Maldives Coastal Protection Project. This program entails the implementation of mangrove afforestation as a means to provide natural barriers against the detrimental effects of coastal erosion and storm surges. The aforementioned aspect has significant importance in the nation's efforts to mitigate the impacts of climate change and safeguard its vulnerable low-lying islands[187].

25.14: CONCLUSION

The presented examples demonstrate the widespread significance and adaptability of mangroves in tackling many environmental and developmental obstacles. Mangroves show their versatility in many settings, including urban development, coastal

[186] National Parks Board, Singapore. (2014). Mandai Mangrove and Mudflats. https://www.nparks.gov.sg/gardens-parks-and-nature/mandai-mangrove-and-mudflats

[187] United Nations Development Programme. (2009). Maldives Coastal Protection Project. https://www.mv.undp.org/content/maldives/en/home/projects/maldives-coastal-protection.html

defense, sustainable tourism, and their contribution to fostering resilient and sustainable island communities.

Chapter 26:

CHALLENGES FACED AND LESSONS LEARNED

In the course of our investigation into initiatives involving the establishment and rehabilitation of islands via the use of mangroves, it is essential to recognize the obstacles faced and the significant insights gained along the process. Although these programs have shown the ability of mangroves to contribute significantly to sustainable development, they have encountered several challenges. This chapter explores the typical difficulties encountered by such initiatives and the valuable insights gained from successfully addressing them.

26.1: COMMON CHALLENGES IN MANGROVE-BASED ISLAND BUILDING AND RESTORATION

26.1.1: Funding and Resource Constraints: Securing sufficient financing and resources to support mangrove restoration initiatives emerges as a prominent concern. The establishment and maintenance of mangrove nurseries, the conduct of scientific research, and the engagement of local populations need financial commitments that may not always be easily accessible[188].

Lesson Learned: The use of inventive financial strategies, collaborations with both governmental and non-governmental entities, and the incorporation of mangrove restoration into wider environmental endeavors may effectively address the obstacles related to finance.

26.1.2: Community Engagement and Livelihoods: The process of involving local populations in the restoration of mangroves might present inherent complexities. The reconciliation of conservation objectives with the socio-economic interests of communities may give rise to disputes. Moreover, the task of identifying sustainable livelihood options for populations who rely on mangrove-detrimental practices, such as shrimp aquaculture or timber harvesting, may present significant difficulties[189].

---

[188] Padhy, S. R., P. K. Dash, and P. Bhattacharyya. "Challenges, opportunities, and climate change adaptation strategies of mangrove-agriculture ecosystem in the Sundarbans, India: A review." *Wetlands Ecology and Management* (2022): 1-16.

[189] Ranjan, Ram. "Optimal mangrove restoration through community engagement on coastal lands facing climatic risks: The case of Sundarbans region in India." *Land Use Policy* 81 (2019): 736-749.

Lesson Learned: The establishment of trust may be facilitated by engaging in inclusive involvement, providing alternative livelihood alternatives, and highlighting the concrete advantages associated with mangrove restoration, such as the augmentation of fish populations. This approach has the potential to cultivate support within the community.

26.1.3: Climate Change and Sea-Level Rise: The rising sea levels and the heightened frequency and severity of storms resulting from climate change present substantial risks to mangrove ecosystems. The aforementioned alterations have the potential to result in the mortality of mangroves and the erosion of coastal areas, hence compromising efforts aimed at restoration[190].

Lesson Learned: The use of resilient architectural designs and the cultivation of climate-resilient mangrove species may augment the capacity of mangrove ecosystems to endure and mitigate the consequences of climate change.

26.1.4: Invasive Species and Disease: The detrimental impact of invasive species, shown by the Rhizophora scale insect, and diseases, such as the Ceratocystis manginecans fungus, on mangrove forests is profound. The task of controlling and

---

[190] Ellison, Joanna C. "How South Pacific mangroves may respond to predicted climate change and sea-level rise." *Climate change in the South Pacific: impacts and responses in Australia, New Zealand, and small island states* (2000): 289-300.

managing these dangers might provide significant challenges.

Lesson Learned: The implementation of timely identification and prompt intervention, in conjunction with the enforcement of isolation protocols, may effectively mitigate the spread of invasive species and illnesses.

26.1.5: Monitoring and Maintenance: The implementation of robust monitoring and sustained maintenance practices is crucial for ensuring the efficacy and longevity of mangrove restoration initiatives. Consistent efforts are necessary to ensure the life and development of planted mangroves.

Lesson Learned: The successful continuation of mangrove restoration initiatives relies on the effective implementation of resilient monitoring methods, active involvement of local people in maintenance operations, and the integration of adaptive management measures.

26.2: LESSONS LEARNED FROM REAL-WORLD EXAMPLES

Now, let us analyze the insights gained from the empirical instances of mangrove-based island construction and restoration initiatives that were previously investigated.

Integrated Approaches Yield Success: Integrated techniques that combine ecological restoration with socio-economic development are often used in successful projects. By effectively catering to the requirements of both ecosystems and communities, these programs are able to attain more substantial and enduring outcomes.

Community Ownership and Empowerment: Projects that place community participation as a priority, empower local individuals, and demonstrate respect for traditional knowledge and customs are more likely to get robust support. The inclusion of communities in decision-making processes and project execution has the potential to significantly increase the likelihood of achieving successful outcomes.

Adaptation to Local Conditions: It is important to acknowledge the distinct attributes and difficulties associated with each location. Projects that adapt their techniques to the specific characteristics of the local environment, including factors such as hydrology, soil quality, and climate, are more likely to achieve success and flourish.

Flexibility and Adaptation: The field of mangrove restoration is continuously developing. Projects that exhibit flexibility, use adaptive management approaches, and modify their tactics in accordance with continuous monitoring and research are more proficient in addressing dynamic circumstances.

Scientific Research and Data-Driven Decisions: The use of science-based methodologies, such as rigorous research and monitoring, plays a pivotal role in shaping decision-making processes and enhancing the efficacy of restoration initiatives. The availability of accurate and dependable data is of utmost importance in evaluating advancements and making well-informed modifications.

Policy and Legal Frameworks: Projects that are in accordance with national and international rules, regulations, and frameworks pertaining to the preservation of biodiversity and the adaptation to climate change are more prone to obtaining funding and attaining long-term sustainability.

Global Collaboration: The enhancement of mangrove restoration efforts is facilitated via collaboration among many stakeholders, including governments, non-governmental organizations, local communities, and foreign partners. This collaborative approach contributes to the increased effect and scalability of such restoration initiatives. The dissemination of information and resources at a worldwide level is of utmost importance.

26.3: CONCLUSION

The obstacles encountered in mangrove-based island construction and restoration initiatives are multifaceted and intricate. Nevertheless, the experiences gained from these obstacles provide significant knowledge about the strategies for effectively managing ecological restoration, engaging with communities, and enhancing climate resilience. By implementing these teachings, forthcoming endeavors might capitalize on the achievements of their forerunners and make valuable contributions towards the sustainable development of coastal regions abundant in mangroves.

Chapter 27:
TIMEFRAMES OF MANGROVE ISLAND HABITAT RESTORATION PROJECTS

Restoration initiatives aimed at rehabilitating mangrove island habitats are multifaceted endeavors that need thorough planning, precise execution, and a considerable degree of patience. The duration of these projects might exhibit substantial variation depending on variables such as the magnitude of the project, its geographical location, the availability of financial resources, and the prevailing ecological circumstances. This chapter delves into the temporal dimensions of restoring mangrove island habitats, including the whole process from project inception to ensuring long-term viability. Our analysis is informed by empirical evidence derived from practical case studies and scholarly research.

27.1: PROJECT INITIATION AND PLANNING PHASE

The initiation of a mangrove island habitat restoration project starts with the phase of conception and planning. During this phase, project leaders and stakeholders establish the fundamental framework for the whole undertaking[191].

27.1.1: Setting Clear Objectives and Goals Clearly stated objectives and goals serve as a strategic guide for the project, ensuring that all endeavors are in accordance with the desired results. During the planning phase, it is essential to allocate a significant amount of work towards the clarification and documentation of these goals.

27.1.2: Site Selection and Feasibility Assessment: In order to identify appropriate locations for restoration efforts, it is essential to conduct a comprehensive feasibility evaluation that encompasses biological, hydrological, and socio-economic considerations. Hastening the progression of this particular stage may result in unanticipated difficulties in the future.

27.1.3: Establishing Partnerships and Securing Funding: The establishment of collaborations with governmental agencies, non-governmental groups, and local communities is a labor-intensive but necessary endeavor. Acquiring financial resources, whether via grants, contributions, or public-private partnerships, may be a time-consuming process.

---

[191] Ellison, Aaron M. "Mangrove restoration: do we know enough?." *Restoration ecology* 8, no. 3 (2000): 219-229.

27.2: Preparatory Phase: Preparing the Island Environment

Prior to the planting of mangroves, it is necessary to carry out preliminary actions at the restoration site. The implementation of this phase is crucial in establishing an appropriate environment conducive to the development of mangroves[192].

27.2.1: Site Preparation and Soil Improvement: Sufficient time should be allowed for the preparation of the site, which involves tasks such as the removal of invasive plants and the improvement of soil quality. Hastening the progression of this stage may potentially jeopardize the project's overall efficacy and sustainability.

27.2.2: Water Management Infrastructure: The design and implementation of efficient water management infrastructure, such as tidal channels and bunds, play a pivotal role in establishing appropriate hydrological conditions for mangroves.

27.2.3: Nursery Establishment and Seed Collection: The process of establishing nurseries and harvesting mangrove seeds requires careful and systematic preparation. Hasty endeavors may lead to

---

[192] Paul, Ashis Kr, Ratnadip Ray, Amrit Kamila, and Subrata Jana. "Mangrove degradation in the Sundarbans." *Coastal wetlands: alteration and remediation* (2017): 357-392.

substandard seed quality and restricted genetic variability.

27.3: PLANTING AND GROWTH CARE PHASE

After the site has been adequately prepared, the process of planting mangroves can commence. Nevertheless, the achievement and expansion of a venture necessitate continuous attention and upkeep.

27.3.1: Planting Techniques and Timing: The selection of planting methods and the appropriate period for planting should be in accordance with the specific ecological parameters and meteorological patterns of the local area. Planting that is done hastily or at an inappropriate time might result in elevated rates of death.

27.3.2: Post-Planting Care and Monitoring: Continuous monitoring and management are crucial for the long-term viability and expansion of mangrove ecosystems. Failure to consider this step might lead to delays and diminished project outcomes.

27.4: LONG-TERM SUSTAINABILITY PHASE

The restoration of sustainable mangrove island habitats extends beyond the first phase of growth, since it involves the broader objective of guaranteeing the long-term resilience and durability of the rehabilitated ecosystem[193].

27.4.1: Community Engagement and Socio-Economic Integration: The establishment and maintenance of enduring connections with local communities, as well as the incorporation of sustainable livelihoods, are continuous endeavors. Failure to consider these factors might place a burden on community assistance.

27.4.2: Climate Resilience and Adaptation: It is essential for mangrove restoration initiatives to maintain a high degree of adaptability in response to evolving environmental circumstances, particularly in light of the challenges posed by climate change. Ongoing research and adaption are necessary.

27.4.3: Legal Protection and Governance

The establishment and enforcement of legislative safeguards for mangrove ecosystems need ongoing and consistent efforts. It is essential that these endeavors be in accordance with both national and international conservation frameworks.

---

[193] Santos, Luciana Cavalcanti Maia, Marília Cunha-Lignon, Yara Schaeffer-Novelli, and Gilberto Cintrón-Molero. "Long-term effects of oil pollution in mangrove forests (Baixada Santista, Southeast Brazil) detected using a GIS-based multitemporal analysis of aerial photographs." *Brazilian Journal of Oceanography* 60 (2012): 159-170.

27.4.4. Research and Innovation: Continual research and innovation play a crucial role in improving project results and disseminating information to the wider restoration community.

27.5. CONCLUSION: THE CONTINUUM OF MANGROVE ISLAND HABITAT RESTORATION

Restoration initiatives focused on mangrove island habitats are not limited to certain timeframes, but instead include a continuous process that extends over many years, and sometimes even decades. The acknowledgment of the temporal intricacies associated with these undertakings is crucial for their overall success. By drawing lessons from previous experiences, sticking to established guidelines, and maintaining flexibility, individuals may effectively traverse the complex timescales associated with mangrove restoration efforts and make meaningful contributions to the long-term viability of these crucial coastal ecosystems.

PART X: SUSTAINABILITY AND FUTURE PROSPECTS

Chapter 28:

LONG-TERM SUSTAINABILITY OF MANGROVE-BASED ISLAND DWELLINGS

Sustainability is more than rhetoric—it guides our planet's long-term well-being. In mangrove-based island villages, sustainability is multifaceted. This function addresses ecological, social, economic, and architectural challenges. This chapter discusses the methods used to preserve these island settings. We examine the long-term sustainability of island habitations constructed on mangrove ecosystems using actual facts, creative architectural principles, and academic research.

28.1: ECOLOGICAL SUSTAINABILITY

28.1.1: Maintaining Ecosystem Health: Long-term ecological viability requires mangrove ecosystem maintenance. According to Das, Sanyal, and Basak (2018), the Sundarbans Biosphere Reserve is a good example. As a wetland ecosystem, the mangrove forest requires regular monitoring and assessment to safeguard against environmental pressures. These findings are important for mangrove conservation and sustainability, especially given growing issues. In 2018, Das, Sanyal, and Basak studied the issue. Sundarbans mangroves are remarkable wetlands[194].

Lesson Learned: The maintenance of long-term sustainability necessitates the ongoing surveillance of the health and resilience of the mangrove ecosystem in response to environmental stresses.

28.1.2: Biodiversity Conservation: Bonaire's Mangrove Restoration: The research conducted by Valiela, Bowen, and York (2001) examines the instance of Bonaire's Mangrove Restoration, emphasizing the crucial importance of mangrove forests in the conservation of world biodiversity. The ecosystems under question, which are considered to be one of the most endangered tropical settings globally, have a crucial function in preserving a diverse range of species. The study conducted by Valiela et al. emphasizes the pressing need of implementing comprehensive conservation initiatives in order to save these crucial ecosystems and the wide array of organisms they sustain[195].

---

[194] Das, S., Sanyal, S., & Basak, J. K. (2018). The Sundarbans mangrove forest: A unique wetland ecosystem. Springer.

[195] Valiela, I., Bowen, J. L., & York, J. K. (2001). Mangrove forests:

Lesson Learned: The long-term ecological health of the mangrove ecosystem relies heavily on the preservation of biodiversity within it.

28.2: SOCIAL SUSTAINABILITY

28.2.1: Community Engagement and Livelihoods: The study conducted by Cinner et al. (2016) on the global coral reefs highlights the instance of Kuna Yala's Community-Managed Mangroves, which exemplifies effective community involvement and integration of livelihoods. The active engagement of the Kuna Yala community in the management of mangrove ecosystems serves as a noteworthy example of the possibilities for sustainable cohabitation with these crucial habitats. Moreover, it underscores the favorable outcomes that such endeavors may provide for the well-being of local communities. The research conducted by Cinner et al. highlights the significance of community-led conservation efforts in attaining social sustainability in coastal areas[196].

One of the world's threatened major tropical environments. BioScience, 51(10), 807-815.

[196] Cinner, J. E., Huchery, C., MacNeil, M. A., Graham, N. A., McClanahan, T. R., Maina, J., & Maire, E. (2016). Bright spots among the world's coral reefs. Nature, 535(7612), 416-419.

Lesson Learned: The inclusion of local people and the provision of sustainable livelihood opportunities are fundamental components of social sustainability.

28.2.2: Cultural Preservation: Smith et al. (2018) conducted a case study on Indigenous Mangrove protection in Australia, examining the relationship between Indigenous peoples and the protection of mangroves. This work offers significant contributions to the understanding of how cultural preservation and ecological sustainability interact. The significance of recognizing the profound cultural connections between Indigenous populations and mangrove ecosystems is emphasized by the study. By maintaining and safeguarding these interconnections, communities have the potential to not only enhance social cohesiveness but also provide the groundwork for a sustained dedication to the preservation of mangroves. The research conducted by Smith et al. emphasizes the significance of cultural factors in the attainment of social sustainability within coastal areas[197].

Lesson Learned: The acknowledgment and safeguarding of the cultural importance of mangroves contribute to the promotion of societal unity and the establishment of enduring dedication towards their preservation.

[197] Smith, T. F., Smith, T. M., & Harrison, S. R. (2018). Indigenous peoples and mangrove conservation: Experiences in Australia and overseas. Australian Geographer, 49(4), 519-536.

28.3: ECONOMIC SUSTAINABILITY

28.3.1: Tulum's Eco-Tourism Development: The research conducted by Zafra-Calvo et al. (2014) examines the economic assessment of ecosystem services in the context of Tulum's Eco-Tourism Development. This case study provides significant insights into the relationship between economic sustainability and the protection of mangroves. This research highlights the importance of sustainable tourism and eco-economic models in both generating money and making substantial contributions to the long-term protection of mangrove ecosystems. Communities may generate a sustainable source of income while also maintaining the biological integrity of mangroves by showcasing the economic worth of these natural areas. The research conducted by Zafra-Calvo et al. highlights the significance of adopting economic approaches that are in line with the conservation of coastal ecosystems.

Source: Zafra-Calvo, N., Gómez-Baggethun, E., & Ruiz-Pérez, M. (2014). Economic valuation of ecosystem services: The case of the Milluni wetland in the Bolivian Andes. Water, 6(10), 3029-3051.

Lesson Learned: The implementation of sustainable tourism practices and the adoption of eco-economic models have the potential to generate financial resources that can be allocated towards the long-term conservation of mangrove ecosystems.

28.3.2: Sustainable Resource Management: The case study conducted by Snidvongs (2009) examines the sustainable harvesting of mangrove products in Thailand and offers valuable insights into the management of mangroves for sustainable coastal livelihoods. This study serves as an important example in the field of sustainable resource management. The research emphasizes that by implementing meticulous and accountable management strategies for mangrove resources, it is feasible to attain economic sustainability while preserving the ecological integrity of the ecosystem. Communities can maintain a steady income while preserving these vital coastal ecosystems by balancing resource extraction and mangrove habitat conservation. Snidvongs' study emphasizes the need for sustainable resource management to protect communities and mangroves[198].

Lesson Learned: Careful mangrove resource management ensures economic viability and environmental protection.

---

[198] Snidvongs, A. (2009). Managing mangroves for sustainable coastal livelihoods: The case of Thailand. In World forests, society and environment (pp. 193-206). Springer.

28.4: ARCHITECTURAL SUSTAINABILITY

28.4.1. Eco-Friendly Design and Building Materials: Abu-Ghazaleh, Hamze, and Al-Hemdi (2018) use Bali's Green School as a sustainable architecture model. This school exemplifies eco-friendly design and construction materials. This study examines bamboo's building material sustainability, focusing on its low ecological impact and architectural versatility. Sustainable design uses bamboo and other eco-friendly materials to merge with the natural surroundings and lessen its ecological effect at Bali's Green School. The case study emphasizes the need of using eco-friendly building materials and designing for usability and sustainability. This strategy is essential for constructing that benefits the environment and community[199].

Lesson Learned: The field of sustainable architecture incorporates architectural concepts and materials that are environmentally benign, hence minimizing the negative effects on mangrove ecosystems.

28.4.2. Innovative Building Techniques: The case study conducted by Naboni and Edwards (2019) examines the novel building techniques used in the

[199] Abu-Ghazaleh, N., Hamze, M., & Al-Hemdi, S. (2018). Assessing the sustainability of bamboo as a building material. Journal of Building Engineering, 19, 667-679.

construction of Floating Homes in the Netherlands. This research sheds light on a unique architectural approach that prioritizes sustainability. The present research investigates the characteristics of sustainable design, with a specific focus on the significance of adaptable and resilient construction methods, especially in areas susceptible to water-related difficulties. Floating dwellings provide an innovative response to the challenges posed by the increasing sea levels and the vulnerability of metropolitan regions to inundation. Through the implementation of this methodology, communities have the ability to not only effectively tackle environmental issues but also establish pleasant and livable environments that live in peaceful coexistence with bodies of water. This case study highlights the capacity of novel construction strategies to not only decrease environmental effect but also improve the quality of life for occupants. This serves as an illustration of how innovative design may facilitate the progress of environmentally conscious urban growth[200].

Lesson learned: Innovative solutions like floating houses may help solve rising sea levels and mangrove preservation.

---

[200] Naboni, E., & Edwards, B. (2019). The nature of sustainable design. In Nature and Architecture (pp. 83-104). Springer.

28.5: LEGAL AND GOVERNANCE FRAMEWORKS

28.5.1: Legal Protection and Enforcement: The Ramsar Convention Manual (2019) by the Ramsar Convention Secretariat illustrates legal methods for wetland protection. The 1971 international treaty prioritizes wetland protection and appropriate usage, particularly mangrove habitats. The above laws protect these vital ecosystems and promote their responsible usage. The Ramsar Convention is an example of a comprehensive legal framework that recognizes the ecological importance of wetlands and protects them globally. This case study emphasizes the importance of legal processes in safeguarding mangrove ecosystems and the environment[201].

Lesson Learned: Strong legal frameworks and international agreements help safeguard mangroves.

28.5.2: Local Governance and Accountability: McKenna's "Putting it back" investigates Indonesia's local mangrove management case study on local government and accountability. The 2011 Bali, Indonesia case study shows how community-based governance protects coastal ecosystems, including mangroves. This paper presents an examination of the long-standing practice of coastal ecosystem management in Bali, highlighting the significant

[201] Ramsar Convention Secretariat. (2019). The Ramsar convention manual: A guide to the Convention on Wetlands of International Importance (5th ed.). Ramsar Convention Secretariat.

involvement of local populations. Indonesian communities have shown a proficient approach to the management and conservation of mangrove habitats, using collective endeavors and well-established customs. The case study highlights the significance of granting authority to local government systems and engaging people in the process of making decisions. This approach results in heightened responsibility and a lasting dedication to conservation efforts[202].

Lesson Learned: The establishment of efficient local government systems plays a crucial role in promoting accountability and fostering sustained dedication towards conservation endeavors.

28.6: CONCLUSION: THE ROADMAP TO LONG-TERM SUSTAINABILITY

The achievement of long-term sustainability for island residences that are built on mangroves necessitates the adoption of a comprehensive strategy that integrates elements of ecological, social, economic, and architectural sustainability. By studying and drawing insights from successful case studies, and by following to these guiding principles, we may effectively negotiate the many challenges

---

[202] McKenna, A. (2011). "Putting it back": A tradition of coastal ecosystem management in Bali, Indonesia. Coastal Management, 39(2), 129-149.

associated with conserving mangroves, while simultaneously fostering the development of prosperous and sustainable island communities that can endure for future generations.

Chapter 29:

FUTURE TRENDS AND INNOVATIONS IN MANGROVE ISLAND HABITAT RESTORATION

The process of restoring the ecology of mangrove islands is a dynamic endeavor that continually adapts to shifting environmental circumstances, breakthroughs in technology, and novel architectural ideas. This chapter delves into the exploration of future prospects, examining the latest trends and technologies that have the potential to influence forthcoming endeavors in mangrove-based island restoration initiatives. By integrating scientific research, innovative building designs, and progressive policies, we provide a picture of a future whereby mangroves not only endure but flourish within a dynamic global environment.

29.1: Climate-Resilient Design and
Adaptation

29.1.1: Elevated and Floating Structures: Trujillo, Bosire, and Krishna (2020) analyze climate-resilient design and adaptation in The Maldives' Floating Island Project. The research suggests using elevated and floating buildings to counteract increasing sea levels. This concept aims to create artificial floating islands for mangroves and humans. Elevated structures and floating platforms provide a climate-resilient system that adapts to the changing environment. This case study shows how innovative design and adaptability may preserve climate-vulnerable coastal communities[203].

Innovation: Using floating and elevated structures to protect mangroves from rising sea levels and storm surges.

29.1.2. Climate-Adaptive Urban Planning: Ma and Gong (2017) investigate China's Sponge City Initiative and climate-adaptive urban development. This innovative urban planning method addresses climate change's complications. This project uses mangroves as flood absorbers and water filters in metropolitan areas. Planning to integrate mangrove

---

[203] Trujillo, D., Bosire, J., & Krishna, V. V. (2020). Floating islands for habitat restoration and climate change adaptation in the Maldives. Ecological Engineering, 147, 105754.

habitats into cities may help them adapt to climate change. This shows the potential of nature-based urban planning solutions[204].

Innovation: Since mangroves are sponges, incorporating them into urban development may help manage flooding and filter water.

29.2: TECHNOLOGICAL ADVANCEMENTS

29.2.1. Remote Sensing and GIS Applications: Technological Advancements: Alongi (2014)'s case study of satellite-based monitoring in the Sundarbans highlights the use of high-resolution remote sensing and Geographic Information System (GIS) technologies for real-time mangrove health and restoration monitoring. This cutting-edge technology helps researchers and conservationists make educated mangrove ecosystem decisions. These applications are essential for mangrove restoration success[205].

Innovation: Real-time mangrove health and restoration monitoring using high-resolution remote sensing and GIS technologies.

---

[204] Ma, H., & Gong, J. (2017). China's sponge city construction: A policy discussion on water management, infrastructure resilience, and sustainability. Landscape and Urban Planning, 168, 11-18.
[205] Alongi, D. M. (2014). Carbon cycling and storage in mangrove forests. Annual Review of Marine Science, 6, 195-219.

29.2.2: Artificial Intelligence in Restoration Planning: AI-Driven Mangrove Reforestation in Malaysia by Viergever, Mayer, and Reddy (2019) highlights how data analysis and predictive modeling utilizing AI algorithms have improved mangrove preservation and restoration plans. Through AI, this notion enhances repair efficiency and accuracy. To maximize mangrove ecosystem restoration, conservationists use complex algorithms to choose restoration sites. AI might boost mangrove restoration[206].

Innovation: Artificial intelligence (AI) techniques are used in the field of data analysis and predictive modeling to enhance the effectiveness of mangrove restoration initiatives.

29.3: MANGROVE GENETIC RESEARCH

29.3.1: Resilient Mangrove Genotypes: The research conducted by Pil, Boeger, and Muschner (2019) titled "Genetic Diversity in Florida's Mangroves: Investigating Resilient Mangrove Genotypes" provides valuable insights on an essential component of mangrove preservation. This study offers an examination of the genetic diversity and population structure of Avicennia germinans,

[206] Viergever, K. M., Mayer, A. L., & Reddy, S. M. (2019). Using AI to prioritize global mangrove conservation. Nature Communications, 10(1), 1-9.

aiming to contribute to the identification of robust mangrove genotypes. This invention facilitates the focused dissemination of mangrove types that demonstrate increased tolerance to environmental stresses. The use of genetic variety exhibits significant potential in bolstering the resilience of mangrove ecosystems in anticipation of forthcoming adversities[207].

Innovation: locating and spreading the genotypes of mangroves that are more resistant to environmental stresses.

29.4: SUSTAINABLE RESOURCE MANAGEMENT

29.4.1: Carbon Trading and Blue Carbon Initiatives: Carbon Trading and Blue Carbon Initiatives: According to Macreadie, Ewers Lewis, Atwood, Duarte, and Ecosystems (2019), the case study on Blue Carbon Credits in Indonesia is a ground-breaking method of sustainable resource management. In order to establish mangroves as important assets in carbon trading and climate financing, this program takes use of their significant potential to sequester carbon. This invention highlights the economic importance of mangrove habitats in a carbon-conscious society while

[207] Pil, M. W., Boeger, M. R., & Muschner, V. C. (2019). Genetic diversity and population structure of the mangrove tree Avicennia germinans (Acanthaceae) in Florida. PloS One, 14(9), e0222291.

simultaneously fostering their preservation by acknowledging their natural usefulness in reducing climate change[208].

Innovation: Harnessing the carbon sequestration potential of mangroves for carbon trading and climate finance.

29.5: PUBLIC ENGAGEMENT AND EDUCATION

29.5.1. Virtual Reality (VR) Mangrove Experiences: The case study conducted by Wood and May (2019) examines the use of Virtual Reality (VR) technology to enhance public participation and education in mangrove conservation. This innovative technique presents a significant advancement in the field of mangrove education in Thailand. This invention utilizes immersive Virtual Reality (VR) experiences to provide a highly influential learning environment. By providing people with the opportunity to engage with mangrove ecosystems via a virtual platform, this program facilitates the development of a more profound comprehension and admiration for these crucial environments. The use of technology enables active participation of the general people in the

---

[208] Macreadie, P. I., Ewers Lewis, C. J., Atwood, T. B., Duarte, C. M., & Ecosystems, G. (2019). Habitat restoration and carbon: A critical examination of the potential. PLoS Biology, 17(2), e3000365.

pursuit of preserving mangroves, so making a significant contribution towards the overarching objective of ecological sustainability[209].

Innovation: Immersive VR experiences to educate and involve the public in mangrove protection.

29.6. POLICIES AND INTERNATIONAL COOPERATION

29.6.1. The Global Mangrove Watch Initiative: The case study conducted by Hamilton and Casey (2016) highlights the Global Mangrove Watch Initiative as an innovative effort in the field of mangrove conservation. The research focuses on the Global Mangrove Watch Platform, which serves as a prime example of this initiative. The primary objective of this program is to establish a comprehensive worldwide database that offers high-resolution monitoring of mangrove forest coverage during the course of the 21st century. This new technique facilitates worldwide cooperation, allowing for thorough surveillance and study of global mangrove ecosystems. The collaborative endeavors play a crucial role in both the preservation of these vital habitats and the development of

---

[209] Wood, L., & May, M. (2019). Using virtual reality for environmental education: Implications for sustainability. Sustainability, 11(13), 3714.

appropriate policies for their sustainable management[210].

Innovation: The monitoring and protection of mangrove forests across the world is the focus of international cooperation.

29.7. MANGROVE REFORESTATION TECHNOLOGIES

29.7.1. Drone Seeding and Planting: Mangrove Reforestation Technologies: Drone Seeding and Planting Munoz-Rojas et al. (2020) did a case study on drone-based reforestation in Madagascar. To evaluate the regeneration of plants after anthropogenic perturbations, this method utilizes drones and deep learning algorithms. Drones are used for the accurate seeding and planting of mangrove propagules, especially in difficult-to-access, tough, or damaged terrains. This breakthrough greatly aids in the revitalization of these crucial coastal ecosystems by increasing the productivity and efficacy of mangrove replanting projects[211].

---

[210] Hamilton, S. E., & Casey, D. (2016). Creation of a high spatio-temporal resolution global database of continuous mangrove forest cover for the 21st century (CGMFC-21). Global Ecology and Biogeography, 25(6), 729-738.

[211] Munoz-Rojas, M., Abd-Elrahman, A., Boer, M. M., et al. (2020).

Innovation: Precision seeding and planting of mangrove propagules using drones in inaccessible or degraded locations.

29.7.2. Biodegradable Planting Materials: The case study done by Ahmed, Rakib, and Sarker (2021) examines the development of biodegradable planting pots for mangrove regeneration in Bangladesh. This research presents an innovative strategy for achieving sustainable afforestation. The primary objective of this endeavor is to develop biodegradable planting containers that are especially tailored for the purpose of restoring mangrove ecosystems. By using environmentally friendly materials, this methodology effectively mitigates waste generation and minimizes its ecological consequences. This initiative represents a significant advancement in the use of sustainable and ecologically mindful methods for mangrove regeneration[212].

Innovation: The use of sustainable planting materials that effectively reduce waste and mitigate environmental damage.

Drones and deep learning provide cost-effective solutions for assessing vegetation recovery following anthropogenic disturbances. Drones, 4(2), 33.

[212] Ahmed, R., Rakib, M. R., & Sarker, M. A. B. (2021). Development of biodegradable planting pots for mangrove rehabilitation: A sustainable approach in coastal afforestation. Ocean and Coastal Management, 208, 105636.

29.8. NATURAL-BASED SOLUTIONS FOR COASTAL PROTECTION

29.8.1. Mangrove-Based Coastal Defense Systems: Coastal defense systems based on mangroves are natural-based solutions for coastal protection. The scholarly article authored by Gunawardena, Rowan, and Ratnasooriya (2020) presents a case study that examines the integration of mangroves with engineering techniques as a means to strengthen coastal resilience in Sri Lanka. The research showcases a novel method to coastal defense. The focal point of this program is to prioritize the incorporation of mangroves in conjunction with traditional engineering methods in order to enhance the coastal resilience of Sri Lanka. The inclusion of mangrove buffer zones in a planned manner offers improved defense against storm surges and adds to the ecological well-being of coastal regions. The aforementioned approach serves as a prime example of the possibility of using natural-based solutions to enhance the resilience of coastal areas[213].

Innovation: The use of mangroves into coastal engineering endeavors to augment storm surge mitigation.

---

[213] Gunawardena, M., Rowan, J., & Ratnasooriya, W. D. (2020). Integrating mangroves with engineering to enhance coastal resilience in Sri Lanka. Frontiers in Marine Science, 7, 925.

29.9. CITIZEN SCIENCE INITIATIVES

29.9.1. Mangrove Mapping Apps: The importance of involving citizens in mangrove conservation is shown in the case study "MangroveWatch in Australia" conducted by Duke et al. (2007), which focuses on the use of mangrove mapping applications in citizen science initiatives. This effort utilizes mobile apps to facilitate the engagement of residents in the process of mapping and monitoring mangroves. MangroveWatch facilitates the participation of people in the collection and sharing of significant data, so enabling the empowerment of communities and the cultivation of a feeling of responsibility towards these vital ecosystems. The use of citizen science in enhancing mangrove conservation efforts is shown by this novel technique[214].

Innovation: There are mobile applications that provide the active participation of people in the mapping and monitoring of mangrove ecosystems.

---

[214] Duke, N. C., Meynecke, J. O., Dittmann, S., Ellison, A. M., Anger, K., Berger, U., & Saintilan, N. (2007). A world without mangroves? Science, 317(5834), 41-42.

29.10: MANGROVE ECOSYSTEM-BASED TOURISM

29.10.1: Sustainable Mangrove Tourism Models: Sustainable Mangrove Tourism Models Based on the Mangrove Ecosystem: The case study conducted by Honey (2008) examines the concept of sustainable mangrove tourism in Costa Rica. The research emphasizes the potential benefits of eco-friendly tourism in promoting the protection of mangroves and supporting sustainable lifestyles within local communities. The present approach places emphasis on responsible tourist methods that aim to reduce the environmental damage while simultaneously enhancing the advantages for local populations. Costa Rica's approach of incorporating mangrove habitats into the tourist sector serves as a noteworthy demonstration of how tourism may be used as a means for both conservation efforts and socio-economic development[215].

Innovation: Sustainable tourism programs that protect mangrove forests and provide locals with a means of subsistence.

---

[215] Honey, M. (2008). Ecotourism and sustainable development: Who owns paradise? Island Press.

29.11: COLLABORATION WITH INDIGENOUS
KNOWLEDGE

29.11.1: Indigenous Practices in Mangrove Restoration: The case study on "Indigenous Ecological Knowledge in Fiji" by Thaman (2007) highlights the need of using indigenous knowledge and practices towards the restoration of mangrove ecosystems. This fresh strategy recognizes indigenous peoples' extensive knowledge of and cultural investment in their respective ecosystems. Fiji's strategy exemplifies the synergy between indigenous knowledge and long-term success in mangrove restoration by embracing traditional practices for resource management and community development. This synergy not only improves environmental results, but also strengthens cultural fortitude and autonomy in these areas[216].

Innovation: Mangrove restoration efforts may gain cultural and ecological value through incorporating indigenous knowledge and practices.

These cases show the wide range of novel strategies being tested to guarantee the long-term viability and resilience of island communities dependent on mangroves. Learning from these

[216] Thaman, R. R. (2007). Indigenous ecological knowledge and sustainable development in the Pacific Islands: Traditional strategies for resource management and community development. South Pacific Journal of Natural Science, 25(1), 1-18.

advances and modifying them for local conditions is essential for a better long-term relationship with mangrove ecosystems.

29.12: CONCLUSION: SHAPING A RESILIENT MANGROVE FUTURE

Restoration of mangrove island ecosystem is an area ripe for future exploration and development. We may look forward to a future where mangroves not only survive but thrive if we adopt climate-resilient designs, utilize technology, unleash genetic potential, and include the public. Protecting these essential ecosystems for future generations will need international collaboration and innovative strategies. In this chapter, we will look forward to the promising future that lies ahead for mangroves and the people who rely on them.

Chapter 30:

HOW CAN YOU HELP?
CONTRIBUTING TO MANGROVE
ISLAND HABITAT RESTORATION

As we conclude up our investigation of mangrove island habitat restoration, it becomes evident that these amazing ecosystems are not only vital to the well-being of our world, but also promise long-term remedies to some of the most intractable challenges confronting the globe today. The protection of coastal communities, the maintenance of biodiversity, and the mitigation of climate change are just a few of the many important roles that mangroves play. However, they need extensive support to survive and recover. This last part will focus on the potential contributions of various organizations and individuals to the preservation and restoration of mangrove island ecosystems.

30.1: THE POWER OF COLLECTIVE ACTION

The restoration and preservation of mangrove island environments include a diverse array of undertakings, including activities such as the cultivation of mangroves, the undertaking of research initiatives, and the dissemination of knowledge to local populations. Individuals who possess a keen interest in environmental matters, conscientious members of society, or entities seeking to effectuate a beneficial influence have a plethora of avenues via which they might actively engage in this crucial endeavor.

30.2: DONATING TO MANGROVE CONSERVATION ORGANIZATIONS

One effective method to provide assistance to mangrove restoration initiatives is by making monetary donations to well-established conservation groups. These groups often exert significant effort in preserving and rehabilitating mangroves, conducting scientific investigations, and actively involving local populations.

Mangrove Action Project (MAP): The primary objective of the Mangrove Action Project (MAP) is to actively engage in the conservation and rehabilitation efforts of mangrove forests on a global scale. Contributions made to MAP provide as financial support for the organization's endeavors in advocacy, research, and educational pursuits[217].

The Nature Conservancy: The international conservation group is actively engaged in various mangrove restoration initiatives across many nations. Donations have the potential to be allocated towards either individual initiatives or the more comprehensive conservation endeavors of the organization[218].

World Wide Fund for Nature (WWF): The World Wildlife Fund (WWF) advocates for the preservation of mangroves as a component of its overarching objective to safeguard biodiversity and ecosystems. Contributions made to the World Wildlife Fund (WWF) play a crucial role in supporting and maintaining their ongoing conservation initiatives.

Website: https://www.worldwildlife.org/initiatives/mangrove-conservation

30.3: VOLUNTEERING YOUR TIME AND SKILLS

Engaging in volunteer work with groups dedicated to mangrove protection may provide significant

---

[217] **Website:** https://mangroveactionproject.org/
[218] **Website:** https://www.nature.org/en-us/what-we-do/our-insights/perspectives/how-to-save-a-mangrove-forest/

personal fulfillment, particularly for those possessing both the necessary skills and the time. Volunteer possibilities include a range of activities, including fieldwork, research, community participation, and several other endeavors.

Volunteer with Local Conservation Groups: Numerous local organizations and community groups are actively engaged in the endeavor of mangrove restoration. Contact these organizations to learn about potential volunteer opportunities.

Scientific Research Expeditions: Certain organizations provide volunteer opportunities for individuals to participate in scientific expeditions that are specifically dedicated to the investigation and restoration of mangroves. These encounters provide a unique opportunity to make contributions to existing initiatives[219].

30.4: CONTRIBUTING MATERIALS AND RESOURCES

Providing crucial materials or resources as donations might be of significant value in supporting mangrove restoration operations. This may include the provision of various resources such as equipment, propagules, or materials for the purpose of creating boardwalks and educational facilities inside mangrove ecosystems.

---

[219] https://www.opwall.com/

Material Donations: It is recommended to establish communication with nearby mangrove restoration initiatives in order to ascertain their unique requirements. Frequently, the acquisition of objects such as horticultural implements, watercraft, or protective equipment is necessary.

Planting Propagules: Mangrove propagule donations are appreciated by several groups. These are often in great demand and may be utilized in forestry projects.

30.5: SHARING YOUR DESIGN AND ENGINEERING SKILLS

Professionals in the fields of architecture, engineering, and design possess the necessary knowledge to greatly impact the restoration of sustainable mangrove island habitats. Engage in partnerships with various groups to foster the creation of cutting-edge and environmentally sustainable infrastructure inside mangrove ecosystems.

Eco-Architecture Design: Collaborate with conservation groups in order to develop environmentally sustainable designs for visitor centers, boardwalks, and observation platforms that are compatible with mangrove ecosystems.

Engineering Solutions: Create environmentally friendly seawalls that use mangroves to reduce the effects of storm surge and erosion.

30.6: SPECIFIC VOLUNTEER PROGRAMS AND INITIATIVES

30.6.1: Mangrove Conservation Volunteer Programs: The Ocean Cleanup's "Intercept Project". One might actively engage in volunteering efforts aimed at intercepting and collecting plastic garbage prior to its entry into the seas, including coastal regions that are home to mangrove ecosystems[220].

30.6.2: Earthwatch Institute's "Conserving Mangroves in the Florida Keys": Participate in a scientific research trip aimed at studying and preserving mangroves in the Florida Keys.[221]

30.6.3: Donation and Funding Sources (GlobalGiving): GlobalGiving facilitates the connection between donors and grassroots programs on a global scale, including various endeavors such as efforts focused on the preservation and protection of mangrove ecosystems[222].

---

[220] *Website:* https://theoceancleanup.com/

[221] *Website:* https://earthwatch.org/

[222] *Website:* https://www.globalgiving.org/

30.6.4: Conservation International: Participate in supporting the conservation efforts of Conservation International, namely in safeguarding mangroves and other vital ecosystems[223].

30.7: CONCLUSION

The restoration of mangrove island habitats is a global endeavor that requires the collaborative engagement of people, organizations, and communities on a collective scale. Through active engagement in this endeavor, whether it by means of financial contributions, volunteer work, provision of resources and expertise, or dissemination of one's own abilities, individuals have the capacity to assume a crucial role in safeguarding the ongoing presence and robustness of these extraordinary ecosystems.

Throughout our investigation, it has become evident that mangroves provide viable answers to a multitude of pressing global predicaments. The worth of these efforts is incalculable, since they include a range of important objectives such as limiting climate change, protecting coastal populations, and promoting biodiversity. Active participation has the potential to provide a long-lasting influence, and by collective efforts, we may establish a viable trajectory towards a sustainable

---

[223] *Website:* https://www.conservation.org/

future wherein mangroves flourish, so yielding advantageous outcomes for both the environment and society.

'A sharp, witty and ultimately hopeful analysis of the global baby bust and how each of us can contribute to overcoming it. Morland persuasively demonstrates that a truly feminist society is one where creating life is more frequently chosen'

Anna Rotkirch, Professor in Social Policy and
Women's Studies, University of Helsinki

'The yawning scarcity of children threatens humanity like no other crisis – yet it is scarcely recognized for what it is. Paul Morland makes the case. *No One Left* is a tour-de-force no one can afford to miss'

Catherine Ruth Pakaluk, associate professor at the
Catholic University of America and author of *Hannah's
Children: The Women Quietly Defying the Birth Dearth*

'The fall in fertility rates across the world and particularly in the developed nations has raised serious concerns about whether countries are preparing for its economic implications. As this hugely informative book by Paul Morland explains, unless the trend is sustainably reversed, inward migration can only help for a bit before countries sink into the final stage of the process where the absolute number of people starts to fall'

Vicky Pryce, Chief Economic Adviser, CEBR and former
Joint Head of the UK Government Economic Service

'Essential reading for reckoning with the demographic winter that is now upon us ... Morland has done all of us a great service. We ignore him at our peril'

Erika Bachiochi, Fellow, Ethics and Public Policy Center

ALSO BY PAUL MORLAND

Demographic Engineering:
Population Strategies in Ethnic Conflict

The Human Tide:
How Population Shaped the Modern World

Tomorrow's People:
The Future of Humanity in Ten Numbers

www.paulmorland.co.uk